Sigma 7

The Six Mercury Orbits of Walter M. Schirra, Jr.

Other Springer-Praxis books of related interest by Colin Burgess

NASA's Scientist-Astronauts
with David J. Shayler
2006, ISBN 978-0-387-21897-7

Animals in Space: From Research Rockets to the Space Shuttle
with Chris Dubbs
2007, ISBN 978-0-387-36053-9

The First Soviet Cosmonaut Team: Their Lives, Legacies and Historical Impact
with Rex Hall, M.B.E.
2009, ISBN 978-0-387-84823-5

Selecting the Mercury Seven: The Search for America's First Astronauts
2011, ISBN 978-1-4419-8404-3

Moon Bound: Choosing and Preparing NASA's Lunar Astronauts
2013, ISBN 978-1-4614-3854-0

Freedom 7: The Historic Flight of Alan B. Shepard, Jr.
2014, ISBN 978-3-319-01155-4

Liberty Bell 7: The Suborbital Flight of Virgil I. Grissom
2014, ISBN 978-3-319-04390-6

Friendship 7: The Epic Orbital Flight of John H. Glenn, Jr.
2015, ISBN 978-3-319-15654-5

Aurora 7: The Mercury Space Flight of M. Scott Carpenter
2015, ISBN 978-3-319-20439-0

Interkosmos: The Eastern Bloc's Early Space Program
with Bert Vis
2015, ISBN 978-3-319-24161-6

Colin Burgess

Sigma 7

The Six Mercury Orbits of Walter M. Schirra, Jr.

Published in association with

Chichester, UK

Colin Burgess
Bangor
New South Wales
Australia

SPRINGER-PRAXIS BOOKS IN SPACE EXPLORATION

Springer Praxis Books
ISBN 978-3-319-27982-4 ISBN 978-3-319-27983-1 (eBook)
DOI 10.1007/978-3-319-27983-1

Library of Congress Control Number: 2016939015

Front cover: *Sigma 7* with Wally Schirra still aboard is readied to be hoisted onto the deck of USS *Kearsarge*. (Photo: NASA)

Back cover: *Top left*: The Gemini VI-A crew of Tom Stafford and Wally Schirra. *Top right*: The Apollo 7 crew of Donn Eisele, Schirra and Walt Cunningham. *Bottom*: Schirra with *Sigma 7* artist Cece Bibby (All photos: NASA)

Cover design: Jim Wilkie
Project copy editor: David M. Harland

Printed on acid-free paper

This Springer imprint is published by Springer Nature
The registered company is Springer International Publishing AG Switzerland

Contents

This book is respectfully dedicated to the memory of a remarkable lady:
Josephine Cook ("Jo") Fraser Schirra (1924–2015)

Foreword

Wally Schirra was my friend. It was an honor and a privilege to say that, and to be asked to write the Foreword to this book.

Wally was my boyhood hero – someone I admired and looked up to. If you had asked me my most outrageous dream while I was growing up, it never would have occurred to me that I might meet and then become a close friend of Wally Schirra. Never did I ever dream that I'd genuinely get to know him, laugh with him, and spend countless wonderful hours in his home working with him. While this book will discuss the flight of *Sigma 7*, I want to tell you about my friend.

Growing up in the 1960s, you couldn't help but know the name of Mercury Astronaut Wally Schirra. He was a household name. Back then, when there was a manned launch, everything stopped. Unlike today, 19-inch black and white television sets were wheeled into school classrooms, and we watched each launch with great anticipation. I have vivid memories of the launch of Apollo 7 in 1968, thinking that it took a lot of guts for those three guys to get into that spacecraft after the terrible fire of Apollo 1. At the same time, I knew how cool Wally had been during his aborted Gemini 6 launch, ahead of the first orbital rendezvous with another spacecraft and the playing of *Jingle Bells* on his tiny harmonica during that mission. And, of course, he was an original Mercury Seven astronaut. Wally would say they were "Carpenter, Cooper, Glenn, Grissom, Schirra, Shepard and Slayton: CCGGSSS. I was the 'Smart S.'"

I was first introduced to Wally in the late 1990s. Wally made a pun regarding my last name of Kornfeld – he never missed an opportunity for a pun. I immediately fired back with a pun of my own. He countered, I fired back again. He then said, "You're good! I like you!" I told him that my late father was a similar punster and that I had years and years of practice. A real friendship was born that evening. He and my father would have had fun topping each other's puns.

Speaking of names, we had an astronaut named Wally. Not Walter or Walt – we had Wally. Someone that Mrs. Cleaver would have called to dinner. He stood out from Al, John, Gus, Scott, Gordon and Donald (as NASA called Deke Slayton back then).

Wally and his wife Jo took my wife and me into their personal lives. It was always fun to visit the Schirra home. Jo could be funnier than Wally, and knew how to keep him in check. We went out to dinner or lunch or just gabbed at their kitchen table. There I was, sitting in the home of my boyhood hero. The Schirras, who have been in the company of kings, queens and presidents, became personal friends.

Like most men, Wally loved his toys. I am a scale modeler and had constructed and given him a couple of models for his personal collection. One afternoon, he called to tell me that the Mercury/Atlas model that I had built for him earlier was the exact same scale of a model needed for the expansion of the San Diego Air & Space Museum, which he affectionately referred to as "Wally World." He then told me that they were using this model in their plans and he wanted to know if it was alright with me that he donated it to the museum for future display. I told him that I would be honored to have something that I had built in a museum, but had to add the kicker, "And I get a percentage of the admissions for my donation, right?" Wally roared with laughter, but couldn't resist firing back at me with, "You gave the model to me and I'm donating it, so the kickbacks are all mine!" A typical Wally response.

Wally's laugh was infectious. You knew he was in the room long before you entered it. Once you did enter, you realized that he filled the entire room. As a member of his "inner circle," I became privy to many stories and tales that aren't in any books or magazines and I surely will never tell in public, but it made our friendship that much more special. On a drive from Los Angeles to his home outside of San Diego, I remarked to my friend, Steve, how lucky we were to be invited to lunch with Wally and be so blasé about it when others would have done anything in their power to be in our place. But Wally was our friend and sharing a meal was natural. The next few hours were filled with that infectious laughter. He was truly one of a kind.

While Wally was well known as "Jolly Wally," he also had a very serious, almost stern, side. Wally was a military man and took his job and his opinions seriously. If a topic came up where he had a strong opinion, that laughing twinkle in his eye could quickly turn into the gaze of a fighter pilot. When that look appeared on his face, you could forget about winning that argument. Wally gave many of his opinions in his autobiography, *Schirra's Space*, which was written in 1988. Several times, I asked him if he'd write an online epilogue. He'd laugh and say that no one, especially NASA, would be interested in his opinions anymore. Just the other day, I was listening to the audio book version of *Schirra's Space* and found myself saying that I wish I had won that argument. Wally had a lot to say, both pro and con, about the space program since the book was published. I would be curious to hear what he had to say, today, about the United States depending on the Russians in order to get to the International Space Station. I'm sure he wouldn't be amused. As he often said about the Soyuz space vehicle, "I wouldn't want to fly in one of those dumb things." He was very concerned about safety – and I don't think that he was too happy with Soyuz. His opinion on the commercialization of spaceflight would also make for some very interesting reading, in my opinion. I don't think he'd look fondly on space tourism.

Tracy Kornfeld with Wally Schirra, San Antonio, 2006. (Photo: Tracy Kornfeld)

He always said that space was a dangerous business. Perhaps he'd have a few words to say about people wearing "the black armband" when tourists are involved.

I also had the honor of working with Wally during several astronaut autograph shows where games of one-upmanship between the astronauts were the norm. One year it was "Who can still fit in their flight suit?" Another was the joke that they had to keep building bigger rockets because Wally kept gaining weight. Wally once signed a flight helmet with "The Real Space Cowboy" as a nod to the book that he had co-authored with his friend,

Ed Buckbee, that year, only to be trumped by Soviet cosmonaut Alexei Leonov, who signed the same helmet with "The Real Siberian Space Cowboy." You never knew what would happen at those "do not miss" events. Even something as simple as sitting between Wally and fellow Mercury astronaut Gordon Cooper, watching them try to steal the one silver Sharpie pen that they shared became a hilarious game. It was non-stop entertainment.

When my wife received a terminal cancer diagnosis, Wally immediately told me that we needed to get away and have some down time, and offered the use of his home on Kauai, Hawaii for as long as we needed. When he assured me that this wasn't one of his classic "gotchas," we were able to spend ten wonderful days at one of the most beautiful places on Earth. My friend was extremely generous. After my wife's passing, he sent me a handwritten letter of condolence saying, in part that "there is no solution to health problems if destiny wins." These words came back to me after hearing of his sudden passing in 2007 at the age of 84. I still have this handwritten note under lock and key because it's so special to me.

I miss Wally so much. I miss his puns, his jokes, his laughter, the unexpected phone calls, the e-mails filled with jokes that made you groan; drinking KJ with him at the bar, how funny it was when he'd answer other people's cell phones with "Schirra here," and of course trying to catch him on a "turtle" joke. But mostly, I'll miss the man himself.

I was also Wally's webmaster. Though he couldn't understand why people wanted to read more about him than was already in print, I'll keep the dream alive at *www.wallyschirra.com* for as long as his family wants me to keep it going. His site had one million visitors in the first 36 hours after he passed away. I will remember how I felt during his memorial service when the fighter jets flew over the cemetery in the missing-man formation, and the emptiness that I truly felt a year after also saying goodbye to my wife.

The laughter at his memorial service was perfect for the man known as "Jolly Wally." Several of us had microphone time. Emcee Mark Larson spoke at length of Wally's "groaner" jokes and how his home number was still listed in the phone book. When comedian Bill ("José Jiménez") Dana took the microphone to speak, the battery died and no one could hear him. Bill just looked straight up and started shaking his fist. Another Wally "gotcha" to remember him by. A garage door-opener battery solved the problem. Bill noted that he needed a "clicker" to talk about Wally. After the service, several of us sat in former astronaut nurse Dee O'Hara's hotel room and told Wally stories into the wee hours. There were so many stories. We all adored him. A solemn event turned around by a remarkable personality.

In conclusion, I will say what I always said to him when we parted company: "Keep your feet dry, Captain, until the next time we meet. Smooth sailing and fair winds." I'd snap a salute, which he would return and then listen to his laughter as he drove away.

We also lost Wally's wonderful wife, Jo, on April 27, 2015 at the age of 91. I was able to send a first draft of this Foreword to the Schirra family several weeks before Jo's passing. I am thankful to have received their blessings for what I have written and for the friendship that continues with the Schirra family to this day.

Tracy Kornfeld
Ridgefield, Connecticut

Acknowledgements

There are no set rules about writing acknowledgements. Every author does them differently, and I normally adhere to my standard practice of listing everyone who has helped me to any degree in all of my books. In putting together this book, however, I realized that my very first published book – on the Australian POW experience – was released some 30 years ago, in 1985. I would therefore like to express my overall and sincere thanks to the countless people who have assisted me over those three decades, for their recollections, guidance, support, kind words, and even their loyalty. I am very humbly grateful to each and every one of you, and I offer my deepest thanks for making these years so unforgettable and mostly enjoyable.

In compiling this book, I must thank first and foremost some good people at the San Diego Air & Space Museum for allowing me access to their files on Wally Schirra, a beloved patron who had a passion for the museum, which in typical Wally fashion he referred to as "Wally's World." To long-time friend and prodigious writing collaborator Francis French, Director of Education at the museum; to Director of Library and Archives, Katrina Pescador; Assistant Archivist Debbie Seracini; aviation historian Gordon Permann; and the education department's Rossco Davis and Shalene Baxter, my enduring thanks for all that you have done to aid me in realizing this book. In recognition of their kind and ongoing assistance, half of the proceeds from this publication will go to support the future work of the museum library in Wally's name.

Many thanks also to Suzanne Schirra for her help and support during the busy and sad time following the passing of her beloved mother, Jo. To Suzanne and her brother Marty my condolences and very best regards.

Additional thanks go to those who were close to the *Sigma 7* story and who so willingly gave their time and memories to aid the story. They are: Col. Kermit ("Andy") and Martha Andrus, Elwood Johnson, Dee O'Hara, Bruce and Nonnie Owens, Alan Rochford, John Stonesifer, and Shirley Sineath Watson. Assistance, support and encouragement also came from: George Carter (Oradell Borough Archivist), Steve Hankow (*Farthest Reaches*), Hunter Hollins and Michael Neufeld (Space History Department, National Air & Space Museum), Richard Jurek, Tracy Kornfeld, Bruce Moody, Linda Pabian (Oradell Public Library),

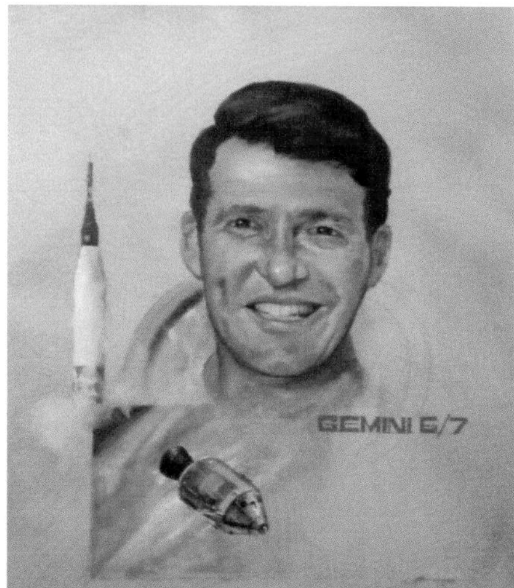

The two portraits of Wally Schirra by artist Craig Kodera that are on display at the San Diego Air & Space Museum, California. (Images and permission courtesy of San Diego Air & Space Museum/Gordon Permann)

J.L. Pickering (*Retro Space Images*), Janice Campbell Pierce, Eddie Pugh, Scott Sacknoff (*Quest: The History of Spaceflight Quarterly*), David Meerman Scott, and Elijah Smith.

A bouquet of thanks as always go to Clive Horwood and his Praxis team in the U.K. for their ongoing enthusiasm for my work, as well as the encouragement and support of Maury Solomon, Editor of Physics and Astronomy, and Assistant Editor Nora Rawn, both of whom are with Springer in New York. And acknowledgements once again go to Jim Wilkie for his outstanding cover artwork and to the master copyeditor and author who always manages to weed out embarrassing errors, sort out my scrambled syntax, and provide that critical final polish to my work, my friend, David M. Harland.

Author's prologue

My first encounter with the truly heroic NASA astronaut Capt. Wally Schirra took place on Sunday, 13 March 1966. I was 19 years old and fully swept up in the romance, excitement, and drama of space travel and astronauts.

Along with fellow astronaut Col. Frank Borman and their wives, Wally was on a daunting three-week goodwill tour of eight countries following their historic December 1965 orbital rendezvous as commanders of Gemini VII and Gemini VI-A respectively. After Japan, Korea, Formosa, Malaysia, Thailand, and the Philippines, they arrived in Australia. The venue for their only fully public appearance in Sydney (in fact their only day in Sydney) was at the Roselands Shopping Centre in nearby Wiley Park. I was part of the audience on that memorable day.

A stage had been prepared and Frank Borman spoke first, thanking everyone for turning up and saying how much they were enjoying their all-too-brief stay in Sydney. Wally Schirra then stepped up to the microphone, saying (with a broad smile on his face) that he admired everyone for coming to see them on a sunny Sunday instead of going to "Bond-ee Beach," and that everyone must be looking forward to their "styk 'n' ex" for lunch. After a few more words and a rousing "We love Australia!" the speeches were over. There was thunderous applause, and the two astronauts gallantly stepped into the 4,000-strong crowd to shake a few hands. I really wanted to meet Wally, but he was facing the wrong way as they passed by in the crush of people. However I still managed to shake the hand of Frank Borman.

The next time I saw Wally was at the Kennedy Space Center on Thursday morning, 29 October 1998. I was there at the kind invitation of crewmember Scott Parazynski to witness the launch of shuttle mission STS-95, carrying Senator John Glenn back into orbit. It was early morning and I noticed Wally standing with Gene Cernan near a press tent, so I walked over and introduced myself. I told Wally about hearing his brief talk in Sydney 32 years earlier, and repeated what he said on that occasion. Wally roared with laughter and we chatted for a few more precious minutes before he was called to reporting duties.

The next occasion was in San Diego on Saturday, 18 January 2003. Francis French and I had just concluded an interview with astronauts' nurse Dee O'Hara, and Francis had to

As Frank Borman and Wally Schirra make their way through the crowd, I am standing towards the back (arrowed), but I still managed to shake Borman's hand as they swept by. (Photo: Lloyd Bott, Australian Department of Supply)

leave for another engagement. It was getting late, so Dee asked what we should do about dinner, and then said, "I'll just give Wally and Jo a call and see if they'd like to join us." Happily they were available and we all enjoyed a night that was filled with recollections and laughter. It was during dinner that Wally laughingly bestowed upon me the nickname "Semi-Colon." From then on, any time he saw me at a space event in the States he would greet me with, "Hey there, Semi-Colon!" It made me feel ten feet tall to be so recognized, and with such friendship, by a man I had always regarded as a great and lovable hero.

It was hard not to like or admire Wally Schirra. He flew an almost technically perfect Mercury flight aboard his *Sigma 7* spacecraft in 1962. Then in 1965 he saved the Gemini VI-A mission by not initiating a launch pad abort when the Titan II rocket failed to lift off after ignition. A few days later he and Tom Stafford flew an historic joint mission with the Gemini VII spacecraft. As commander of the backup crew for the first Apollo mission, Wally and his two crewmembers had to take over that role after the launch pad tragedy that

The last time I saw Wally Schirra was at an autograph show in San Antonio, Texas, in 2006. Seated around a hotel breakfast table was a very impressive gathering of space folks. From *left*: Colin Burgess, Francis French, Dee O'Hara, Wally Schirra, Erin French, Jeannie Bassett, Ed Buckbee, and Cece Bibby. (Photo: Author's collection)

claimed the lives of three colleagues in January 1967. By completing the Apollo 7 mission in 1968, Wally became the first (and only) Mercury astronaut to fly on all three NASA programs to that time: Mercury, Gemini and Apollo. Throughout his time with NASA he exuded a calm and thorough professionalism, but even when things were at their toughest Wally would be the first one to break the ice with a joke or a pun, or even carry out one of his infamous "gotchas" – which will be further explained in this book.

Wally Schirra's passing on 3 May 2007 came as a dreadful shock to me, as it did to so many others who knew him. It was hard to realize that he would never be present in his Hawaiian shirt at any future space shows. He always commanded a delighted audience wherever he went, and his booming voice and frequent laugh would often fill the most cavernous of rooms. Although we miss him a great deal, we are richer for the fond and enduring memories that he left with us as his legacy.

Ave atque vale, Wally Schirra

Wally Schirra sent me this signed photo as a souvenir of his Gemini VI-A flight with Tom Stafford. It was another of his famous "gotchas" that I truly treasure. (Photo: NASA)

1

A pilot born of pilots

"You don't raise heroes; you raise sons. And if you treat them like sons they'll turn out to be heroes, even if it's just in your eyes."

Walter M. Schirra. Snr. (1893–1973)

They were strong-hearted, incredibly valorous young men. Any pilot who flew into perilous aerial combat in the Great War (as it was known before it became necessary to number them) and lived to tell of his encounters had a valid reason to believe that the gods were on his side. Especially those who took to the air in the AirCo DH.9 single-engine light bomber airplane.

Designed by aviation engineer Geoffrey de Havilland, the DH.9 was introduced into service in 1917, but rapidly gained a poor service reputation over the Western Front, with more aircraft losses attributed to mechanical or performance issues than through enemy action. The engine was notoriously unreliable and underpowered, and there were design problems inherent in the undercarriage, wings, and tail unit. Although tempting fate, pilots and their gunners would often and ruefully refer to the DH.9 as "the flying coffin." And with good cause, because between May and November 1918 two squadrons on the Western Front (Nos. 99 and 104) lost 54 DH.9s shot down, and another 94 written off in accidents.[1]

FLYING ON THE WESTERN FRONT

One such Western Front pilot was 24-year-old Lt. Walter Marty Schirra, who regularly flew a DH.9 on aerial bombing and photo reconnaissance sorties over the war-ravaged land below. It would often strike him as odd to be engaged in shooting down German airplanes or bombing German troops, as his parents, Swiss-born Adam and his wife Josephina Marty Schirra, had only emigrated to the United States from Bavaria in southern Germany some 38 years earlier.

C. Burgess, *Sigma 7*, Springer Praxis Books, DOI 10.1007/978-3-319-27983-1_1

An art collector and talented musician, Adam Schirra had become the one-time principal cornet player with the Philharmonic Society of New York, but later gave up touring to concentrate on teaching the instrument. Like his wife, Adam was of Swiss stock. The Schirra name seems to have originated in Switzerland's Valle Onsernone, and Adam was born in the Italian-speaking village of Loco in Canton Ticino, located between the Swiss Alps in the south, Lake Lucerne to the west, and Lake Zurich to the north.

Philadelphia-born Walt Schirra only became a pilot through a series of mix-ups. An engineering graduate from Columbia University employed at the United Fruit Company in Honduras, he had signed up as a U.S. Army engineer in 1916. Later, however, he heard through the grapevine that only 45 men out of a company of 150 would be commissioned. Determined to take an active part in the war he sought a transfer to the artillery, but to his disappointment this was refused. Then fate lent a hand. As he was about to leave Army headquarters at Fort Myers, Virginia, an adjutant stopped him and asked a question that would dramatically change the course of his life, "How would you like to get in the aviation section of the Signal Corps?"

He quickly signed on, and took a course at the Fort Myers officers' training school as a preliminary to flight training on Curtiss Jenny JN-4 biplanes at Taliaferro Field, near Ft. Worth, Texas. Following flight training, he was posted to a new squadron that was being assembled at Kelly Field, five miles west of San Antonio, Texas. After several months of Army training there, the unit received its official designation as the 28th Aero (Fighter) Squadron, U.S. Air Service (USAS) in June 1917.

As the USAS did not yet have any active front line squadrons, Schirra's squadron was attached to Britain's Royal Flying Corps for training purposes, and was sent to a pilot training facility in Toronto, Canada. The cadets were given lessons in aircraft construction, overhaul, upkeep, motor transport work, aerial gunnery, and other skills. On completing their training, each cadet was commissioned an officer and assigned to the squadron as a first lieutenant, and then on returning to the United States they were told to prepare for overseas embarkation early in 1918.

It was not just flying that had occupied Walt Schirra's thoughts after his squadron was sent to Canada; a few days beforehand he had chanced to meet a pretty young art major and fifth-generation Brooklynite from the Pratt Institute named Florence Shillito Leach, and made up his mind that she was the one and only girl for him.

During breaks from his training, Schirra would send a steady stream of letters to Florence. For her part, she adored the dashing young U.S. Army Signal Corps man, but one letter caught her completely by surprise. "The next thing I got in the mail was an insurance policy for $10,000 made out to me, naming me as his wife. That's how he proposed."[2]

On his return from Canada, there was just one thing Schirra wanted to do before leaving for France. As his astronaut son would later observe, "He married my mother just before catching the troop ship. It was a whirlwind romance."[3] Florence dropped out of art school to get married and the wedding took place on 8 February 1918. Just twelve days later the troop transport ship RMS *Olympic* (White Star Line) left New York Harbor bound for England. Schirra was in his Army Signal Corps uniform, and "he wore boots with spurs," Florence recalled as she waved him goodbye.

Lt. Walter Schirra (standing on left) with two fighter pilot friends, and in civilian attire at right. (Photos courtesy San Diego Air & Space Museum Walter Schirra personal collection)

On arrival in England, the eager recruits completed additional combat training prior to being sent to an airfield in France, located some 35 miles behind the front line. Meanwhile, Schirra's new bride moved to Meriden, Connecticut to live with her husband's family while awaiting his return. Once No. 108 Squadron, Royal Flying Corps, had arrived at their French base, a gunner/observer was assigned to Schirra, to sit immediately behind him in their aircraft. As pilot, Schirra's weapon consisted of a forward-firing Vickers machine gun, whereas his gunner was equipped with a Lewis machine gun mounted on a swiveling scarff ring. Their aircraft could carry up to 460 pounds of bombs below the wings. Back then, however, there was no such lifesaving equipment as a parachute; the unlucky pilot of an aircraft that was shot down would ride his crippled machine down to a generally fatal crash with the ground. If the engine erupted into flames in the air, many doomed pilots would simply leap out of the airplane rather than be roasted alive.

MISSING IN ACTION

Walt Schirra would fly as many as four sorties per day, engaging with enemy aircraft (he is believed to have shot down two airplanes), dropping bombs over German lines, and carrying out hazardous photo reconnaissance missions. His combat flying career came to an abrupt end as was returning from a reconnaissance mission and his aircraft was downed over the French countryside, having been chewed to pieces by shrapnel from enemy ground fire. Thick black smoke belched out of a ruptured crankcase. His gunner, hit by the ground fire, apparently died before the DH.9 landed hard. Schirra, badly injured, was initially transported to a field hospital in Dunkirk.

Walt Schirra in France standing by his AirCo DH.9 bomber aircraft. (Photo: Schirra family)

It was a worrying time for Schirra's family back home when his flow of letters suddenly ceased. Florence's sister-in-law, Alma Schwartz, recalled in 1962 that a letter to her brother had been sent back unopened with "Recipient Deceased" written on the envelope. The return of the letter bearing the terse endorsement came as a complete surprise and was understandably a great shock to the new Mrs. Schirra. And then, to compound their misery, a telegram arrived informing Florence that her husband was missing in action and presumed dead. On returning home after a somber funeral service in a little Catholic church in Meriden, she found another telegram waiting for her, this one bearing the news that her husband was wounded but alive and was being tended in a hospital in Winchester, England. Alma's husband Michael sent an urgent wire to Washington D.C. seeking clarification. They finally learned to their relief that Walt Schirra had indeed survived a serious crash and was being nursed in England.[4]

Much to her distress, Florence would receive two further false alarms by telegram before the war ended. However, after the first of these she refused to believe them. Once peace had been declared, records indicate that Schirra apparently stayed on and served as a pilot for Col. Frank P. Lahm, Chief of the Air Service, Second Army at Toul in

An AirCo DH.9 similar to the airplane flown by Lt. Schirra over France. (Wikipedia Public Domain photo)

north-eastern France, until the unit was dissolved on 15 April 1919. Her husband finally returned home in July 1919.

It was only then that Florence found out that she might have actually lost him for good as a result of misadventure after the war. Walt had admitted to being curious about whether he could fly an S.E.5A biplane through the Arc de Triomphe in Paris, and so one day he roared westwards at low altitude along the Avenue des Champs Élysées towards the monument. In the end he wisely decided not to make the attempt because there were too many power lines in the way, but he nearly gave a lot of people heart failure by trying.[5]

"Dad saw a lot of combat over France shooting down Germans," Wally Schirra wrote in his 1988 memoir *Schirra's Space*, "and on three occasions he was downed and listed as missing in action. His favorite story was about ferrying an aircraft from France to Britain at the end of the war. He asked a French mechanic what was in a box strapped to the fuselage, and the mechanic said it was something called a 'parachute.' He'd flown without a parachute in combat, and when hit, he crashed the plane. Mother held two funerals for him and collected insurance money, which she returned. Dad came home healthy except for a hunk of shrapnel in his leg. To the end of his life he was unable to pass a metal detector test in an airport security check."[6]

After the signing of the Armistice ending the war, Walt Schirra wanted to continue flying but was not interested in holding down a regular job in civil aviation. He even turned down highly paid employment flying the New York-Cleveland route with the U.S. air mail service because that would not provide the freedom of the skies that he desired. The thrill of flying was still there and Walt decided he wanted to fly his own airplane. He finally bought a Curtiss Jenny JN-4D similar to the one in which he had trained at flight school.

The post-war years became renowned as an era of "barnstormers" across the United States. The participants in this dangerous activity were mostly returned war veterans or daredevil pilots who performed aerial stunts to attract a crowd, after which they would sell rides aboard their airplane to those game enough to purchase a ticket. This activity appealed to Walt Schirra and he went on his own barnstorming tour at county fairs around New Jersey, selling airplane rides at a dollar a minute. But Florence had also proved to be a plucky young lady and they decided to form a flying duo. Once the little biplane was airborne at about 65 mph she would clamber out of the rear seat onto the wing and walk along it, holding on to the struts that connected the upper and lower wings. Her husband once explained that the wing walk was just a promotion gimmick designed to drum up business, and help to pay the fuel, oil, and hangar bills. As their son later described:

"In the carefree days before my sister or I was born, my parents had a fine time barnstorming in a Curtiss Jenny. Mom was a wing-walker. With Dad at the controls she would dance on the lower wing of the biplane, using the struts for support. It looked hair-raising and no doubt was. Her act attracted customers who would pay five dollars for a turn around the field. When asked about it later, she would say she gave up wing-walking when I was in the hangar."[7]

As Florence later admitted to a reporter, "Stunt flying at the New Jersey fair was a lot of fun, but we couldn't find enough passengers at one dollar a minute for rides. I never

A wing-walker on a Curtiss Jenny over New Jersey in a similar, dangerous stunt to that performed by Florence Schirra. (Photo: Slideshare.com)

tried to stop my husband from flying, but when Walter was born he decided to sell the plane and go to work."[8]

With the barnstorming venture a financial failure, the Schirras decided that when things went really bad they would seek employment elsewhere. Eventually they rented a home in tree-lined Oradell in northern New Jersey, a small town of around 2,000 people, where Walt had found employment as the borough engineer. Over the next few years the family lived in at least three rented houses, mostly of them on Maple Avenue. There is no record of them ever buying a dwelling, which is why today there is no street marker officially identifying Schirra's boyhood home.

On 12 March 1923 the Schirras celebrated the birth of a son in nearby Hackensack Hospital (Oradell had no hospital of its own) delivered by Dr. George Edwards. The delighted couple named their baby Walter Marty Schirra, Jr. Four years later, in 1927, the arrival of his little sister Georgia Lou completed the family.

GROWING UP IN ORADELL

As a young boy, Wally Schirra attended Sunday School at the local Episcopal Church on Kinderkamack Road, and grew up in a house filled with love and laughter. In 1962, before he flew his Mercury mission, his mother Florence said, "The Schirra family has always had a lot of fun. We kid each other a lot and whenever we get together, there's somebody who gets it. We had a wonderful family life; we lived like a family; we acted like a family. It is wonderful to remember."[9]

His mother would also recall that as a little boy, "All his toys became airplanes. He never used to run his little trucks along the ground. He would pick them up instead, and make a plane sound and fly them through the air. When he was three he was tearing up note paper into dart shapes and flying them across the room."

To assist in raising his young family, Walt Schirra "moonlighted" on various engineering jobs, which included helping to build the northern approach to the Lincoln Tunnel and working on a sewage disposal plant project in Long Island.

In another interview, Florence revealed that Wally started out early as a prankster. "Oh, he was a handful," she said, although adding that he never got into real trouble, but was sometimes so completely mischievous, "that I had to send him to his room for punishment." One time, she remembered, Wally told his playmates that he would show them their school principal dancing for the price of one cent each. After collecting the money, Wally led them into the principal's back yard and pointed out the gentleman's long underwear dancing around in the wind. Dissatisfied by this ruse, his customers complained to Wally's mother and the money was grudgingly refunded.[10]

Flying, it seemed, was always apparent in some way in young Wally's blood. His bedroom was cluttered with airplane models that he had built, and one wall was covered in photographs of his uniformed father as a Great War aviator, along with pictures of the types of aircraft he had flown. Hanging prominently on a nail were his father's soft flying helmet and goggles. Their living room often played host to some of his father's wartime buddies, and young Wally would sit quietly to the side, rapt in the exciting stories they told of aerial combat and other adventures in the air.

An unnamed local Oradell scribe once delved into the 18 years that Wally Schirra lived there before enrolling at the U.S. Naval Academy at Annapolis, Maryland. "He was exactly like any one of several boys in our town. Good boys, intelligent youngsters with normal curiosity about the world in general. Like many other teenagers he was interested in model aircraft and flying, but neither his parents nor his friends felt that this was any particular indication of his future career, for many of his contemporaries were just as expert as he was in model making. And yet possibly this interest did suggest to some people that he might seek to make a career in flying."[11]

Virginia Chapin (later Mrs. William Knight) was assistant to school principal Evelyn Lindstrom as a kindergarten teacher when Wally Schirra began his education in Oradell. She fondly remembered him as a "smiley type" of boy, a little on the chubby side, but a very nice youngster. Helene Mertching, a one-time school superintendent, taught him mathematics for three years. She recalled him as a particularly good student. "He had a sort of shy smile and was most dependable," was her recollection.[12]

On summer breaks, Wally and his sister Georgia would spend many happy days on the farm of their aunt and uncle, Alma and Hungarian-born Michael Schwarz, in the small rural town of Chester on the banks of the Connecticut River. Wally would often be found sticking his head under the hood of his uncle's Model A Ford to see what made it run. He would later own one himself.

This photograph of 371 Maple Street, Oradell, was taken circa 1933–1934, at which time it was occupied by the Schirra family. (Photo: Courtesy George Carter, Oradell Local History Collection)

In 1934, at the qualifying 12 years of age, Schirra joined the Boy Scouts in Oradell as a member of Troop 36. This had become the first scout troop to be organized in the borough two years earlier. He would eventually attain the rank of Scout, First Class. Some later biographies claim that he achieved Eagle Scout, but this was not the case. Oradell historian Frank Vierling wrote in his book *The Delford-Oradell Centennial 1894–1994:* "My boy scouting years were spent in Troop 36 along with Wally Schirra. We had great scout meetings every Friday night in the school gym. We participated in Jamborees with other Troops in the area and had many hiking forays into the countryside. We camped in Saddle River and Paramus."[13]

The late Lew Robinson was the scoutmaster in charge when Wally joined, and during an interview after Schirra's Mercury space flight he said that, like so many other friends and neighbors of the Schirras, he never suspected the heights of achievement to which the boy was destined to soar. "Walter Schirra was a fine lad, very dependable," he ventured. "I'd call him a darned good scout, but, to tell you the truth, I never guessed that he would be one to volunteer for the kind of sky-blazing job he took on – and made good at. I'll say one thing, though; Scout Schirra had the stuff in him to make good at almost anything he tackled. I never found him trying to bluff his way through any merit badge examination. He learned his scout skills thoroughly. And that, I think, was one of the things that helped him get into Annapolis."[14]

One of Wally's schoolboy companions, Herb Landmann, said of his future astronaut friend, "We lived within a block of each other in two or three different houses and we hung around together. We would spend Saturday afternoons in his living room [at 317 Maple Avenue] talking and building model airplanes. I think being a flyer was on his mind. He was a good, natural, intelligent, average type of guy. He was never voted the most likely to succeed. There were a number of people who really impressed me during our school days, but he wasn't one of them. I admire Wally, of course. But I was surprised when they announced the seven original astronauts on TV. I turned to my wife and said, 'By gosh – one of them is Wally!'"[15]

In his youth, Wally Schirra may have harbored a moderate interest in airplanes and flying but he did have other sporting enthusiasms, including ice skating and ice hockey, basketball, soccer, and golf. For a time he even took trumpet lessons, probably at the prompting of his cornet-playing grandfather Adam, but gave up the instrument because it really didn't appeal to him.

Even though it was forbidden by law at the time, he also loved paddling on the nearby Hackensack River in a kayak that he had built in a shop class at the Oradell school. Local historian George Carter recalled for this book that Schirra "would stand on the Oradell Avenue Bridge and create a ruckus so that residents further down would call the police. Wally would be right there, volunteering to go into the water in his kayak to check it out."[16]

On one occasion, Wally's bridge follies caused his parents some grief while Florence was driving across the Oradell Avenue Bridge. She happened to look down in time to see Wally and a friend tip their kayak over and disappear momentarily into the swirling waters. As it transpired the boys were fine, but Wally's mother was so alarmed that she ran off the road and demolished someone's fence.

Another time he and a friend hid their kayak under the bridge, cried out for help and made loud splashing noises, which were so dramatic that two elderly ladies passing by thought someone had fallen or jumped into the water and was experiencing difficulties. As Florence later recalled for a *Life* magazine reporter, the ladies promptly called the police station. Within minutes, squad cars and fire engines had converged on the scene. Shortly after, Wally was back home and innocently resting in his room when the police chief came and knocked on the Schirra's door. He was very stern, and asked Wally to bring his kayak and help search for the body of a drowned person. But the chief was no fool; he knew exactly what had happened. As Florence recalled of those times, "How I used to hate to open the front door and see the police chief again."[17]

When asked years later what he liked most about Oradell, Schirra replied, "The typical American family likes a home among trees, and that's what I liked about Oradell."

As with most boys, Wally always had at least one pet. His last prior to moving on was a German Shepherd named Pepper, which later became a K-9 Corps dog in World War Two and was credited with saving two lives while on sentry duty in Saigon.

In school, Wally was an excellent student, especially in mathematics. "He was good at everything," his mother said. However, he wasn't allowed to go out on school nights unless it was to his Boy Scouts meeting or the library.[18]

Wally graduated from Oradell Junior High School in 1937, where he had been voted the "wittiest student" because of his wild and often ridiculous puns. He often played his prized harmonica, displaying a penchant for swing music. In fact he was one of the promoters of a frenzied harmonica quartet, which gave him an added popularity. He became a member of the school orchestra and other musical groups, and even worked on the class song. Wally was also a reasonably good artist and he contributed to the school's yearbook. As well, he displayed some ability as an actor when he played the leading role in the class play. But his class yearbook could not resist chiding him for his surprising shyness. "Again we are able to remonstrate with one of the shyest of our masculine members. It seems Wally Schirra has a great obsession for gazing at the stage floor. During the play, 'The Promoters,' it was almost impossible to get him to raise his head for identification. Head up, Wally. We don't mind your face."

Wally's ability as a punster also led to the comment that a good class present to him would be an oven fitted with a clock, "to keep Wally's jokes from coming out half-baked." A final comment was somewhat prophetic, saying that he, "Wants to be a West Pointer – ought to be an airplane designer."[19]

A LOVE OF FLYING AND THE GREEN BOWLERS

"Dad still flew during my childhood but just for the fun of it," Schirra would later write in his autobiographical memoir *Schirra's Space*. "He owned an Aeronca C3, and he took me up for the first time when I was eight or nine. I remember we climbed from the runway and headed into a stiff wind – it blew so hard that we were moving backwards in relation to the ground. From that day on I loved flying. Teterboro Airport wasn't far from Oradell, and

I remember riding there on my bicycle to watch the 'airplanes.' It wasn't until later that I learned an aircraft doesn't plane, as a boat does when it moves through water. A wing generates lift passing through air, with the air on top going faster than air underneath, resulting in less pressure on top, thus lift. Aircraft were crafty vehicles, I thought, and I no longer called them airplanes."[20]

One of the main reasons Schirra would pedal the 25 miles to Teterboro Airport, apart from watching whatever aircraft were present, was to talk with a close friend of his father's, Clyde Panghorn, who patiently told the eager youngster all about aircraft and flying. However, his interest in aviation did not truly turn into a passion until one summer's day in 1936, while he was still attending high school. He was riding alongside his father in the two-seater Aeronca (unkindly known to the family as "the bathtub") 3,000 feet above Teterboro when his father nudged him in the arm and shouted, "You take it over!" It was an unforgettable moment for both of them, and the 13-year-old boy's eyes widened with excitement as he clutched the wheel between them and steered the frail craft for 30 minutes, flying over the Palisades and the Hudson River to the fringes of Manhattan. "That was it," his father later recalled. "From then on that boy was hooked on flying."[21]

"My father did not push me into flying," Schirra would recall. "But we were very close, and I had such a complete respect for him that I wanted to be just like him."[22]

Schirra would further his education as a sophomore at Dwight W. Morrow High School in nearby Englewood, where, at that time, Oradell was sending its students. Curiously, given his later achievements, he was never regarded as being particularly outstanding at the school. Retired vice-principal Charles Wildrick served as adviser to the math club, of which Schirra was a member, and he once admitted that he could not recall the youth at all. "He must have been quiet and calm, the A-student type, or I would have remembered him."

Conversely, James Kirkland, who was both vice-principal of the school and coach of the school's soccer and ice hockey team (Schirra played on both), did remember the youth. "He was the kind of fellow who would do a job and not want the limelight. If he played, it was swell. But if he didn't, it was all right with him. In his senior year he was the best fullback we had." Kirkland also said that Schirra was essentially a good team player, and he thought that this contributed a great deal to his later success as an astronaut. He added that Schirra certainly contributed to the success of the soccer team, as they won the league championship in 1938 and again in 1939, and tied with Hackensack High in 1940. He was, however, less successful as a hockey player, missing out on the starting line-up and playing mostly in defense. But the team only lasted two years before it disbanded and hockey was dropped from the senior sports program.

Kirkland was also the school's guidance counselor, and in this capacity he recalled that, "Although he was a good-looking boy, I don't recollect his going out with the girls or even talking about them. He was a thorough gentleman, as were the rest of the group from Oradell that came to school here. There was not much guidance necessary in his case. He knew where he was going. His best subjects were math and science. He wanted to be a civil engineer and follow in his father's footsteps." Kirkland also said that as a good average student, Schirra was always grouped with the better students.[23]

Dwight W. Morrow High School, Englewood. (Photo: Englewood Board of Education)

While attending Morrow High, Schirra won the school's bronze "E." This was awarded for outstanding scholarship and participation in extra-curricular activities, one of which was secretary-treasurer of the boys' cooking club. He was also in the math club and the mask and wig club, and was property manager for the senior play. Kirkland believes Wally may also have been in the astronomy club, although he is not listed as such in the school's year book. He also sang in the school choir for about a year.

His first automobile, which he purchased while still attending Morrow High, was a Ford Model T. He later owned a Model A, and then a 1929 Pontiac convertible. But these cars soon gave way to his pride and joy – a 1932 Plymouth PB, for which he paid the princely sum of $25. Music also played an important part of Schirra's formative years. He and his friends derived great pleasure from attending Frank Dailey's Meadowbrook dance hall in Cedar Grove and listening to live swing performances by such big bands as Harry James, Duke Ellington, the Dorsey Brothers, and Glenn Miller; as well as the young New Jersey-born crooner by the name of Frank Sinatra.

After graduating in 1940, Schirra attended the Newark College of Engineering, one of the oldest and largest professional engineering schools in the United States, where, significantly, he became a member of the Sigma Pi fraternity. Flying lessons continued, but he never got to solo in the Aeronca, because one day his father lent it to a friend who made a brutally hard landing and smashed it to pieces. Schirra was still attending Newark College when Japanese forces attacked Pearl Harbor on Sunday, 7 December 1941, drawing America into World War Two. The next day, students who were in the Reserve Officers Training Corps turned up to class wearing their uniforms, and Schirra said he was "suddenly aware of a national emergency."

The Schirra family was now living at 79 Elizabeth Street in Oradell, but it was time for 18-year-old Wally to move on. With the support of his father, Schirra, eager to become a service pilot, sat for an examination that would qualify him for immediate entrance to the U.S. Military Academy at West Point in New York State, not far from where they lived in northern New Jersey. He realized that a military education would provide him with a quick ticket to a career in aviation. On his exam papers where a preferred military institution was required, he wrote "USMA," because that was where his father wished him to go. He then completed a number of tests in math and science, at the end of which he figured he'd done quite well. Once the exam was finished, the overseer said that anyone wishing to join West Point would be asked to stay behind and complete a further exam on the academy's history. For Schirra, this was a real stopper.

As his future wife Jo would tell the author, "Wally knew that there was way too much history associated with West Point, so he asked if he could get his paper back and changed his preference to USNA – the naval academy. When he was accepted, his father thought there'd been some sort of mistake, but after Wally told him what he'd done he went along with it. I'm glad, because if Wally hadn't changed his mind back then we would never have met."[24]

There was another reason for changing his mind that harked back to his childhood. "There are images which will remain forever and plant a seed in a child's mind," he said. "Walking along the street in a small town, I saw this Navy commander in greens, brown shoes and gold wings." That's class, he said to himself; I want that. The memory of that day was a factor in him changing his mind and going Navy.

In 1942, despite his father's initial disappointment, Schirra was appointed to the Naval Academy by Representative J. Parnell Thomas of the Seventh Congressional District. He entered as a member of the Class of 1946, plunging straight into the curriculum.

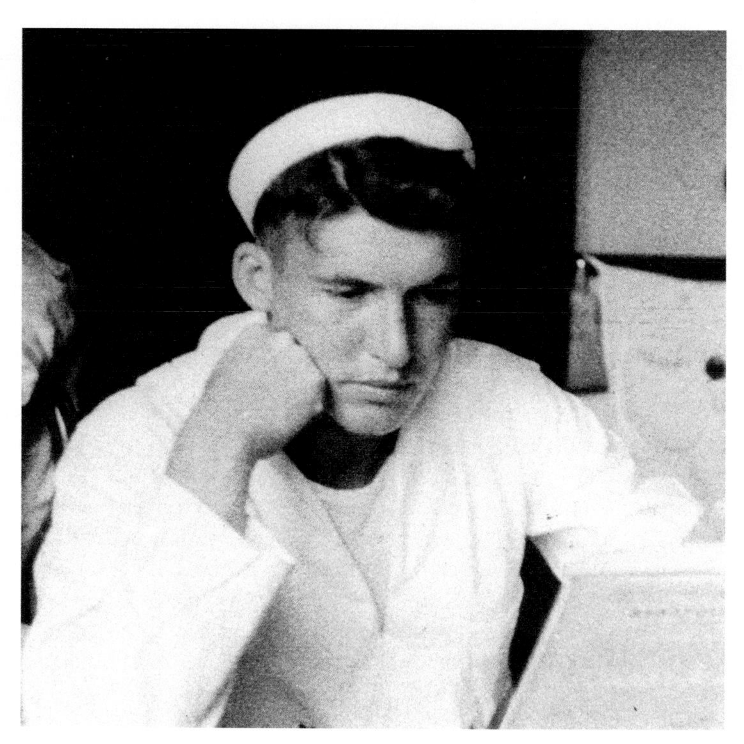

Walter M. Schirra Jr., USNA Class of 1946. (Photo: U.S. Naval Academy)

As the war continued, his class became war-accelerated, cramming a four-year program into three. Midshipman Chief Petty Officer Schirra graduated with a bachelor of science degree on 6 June 1945, just three years after entering the academy. But by that time the war in Europe was over and the war in the Pacific was nearing its climax.

Prior to his Mercury flight, Schirra penned some reflections on his time at the academy, what the training for the first American space flight was like, and which one of the seven Mercury astronauts would be given the honor of making that flight:

U.S. Naval Academy, Annapolis, Maryland. (Photo: Wikipedia/John T. Lowe, Historic American Buildings Survey)

"When I was at the Naval Academy, there was a lot of talk about a secret organization of midshipmen called the Green Bowlers. These were supposed to be the best all-around men at Annapolis, the smartest and nerviest of each class, the men who one day would probably be the top admirals in the Navy. It was said that they wore green bowler hats at their secret meetings. Well, I don't know if there was such a group – and if I knew, I wouldn't say so – but I've always felt proud enough of my profession to think that being a test pilot makes you a Green Bowler in aviation. And for my money the man who takes that first ride in the capsule is going to be the Green Bowler of all time."[25]

After their son graduated from the academy, Schirra's parents moved out of Oradell and briefly relocated to Arlington, Virginia, where he paid them a visit to celebrate. Next, they spent ten years in Japan. Walt was a consulting engineer for the Pacific Air Command, and Florence taught English to students at Keio University in Tokyo. After that was six years at Foster Village in Hawaii, where Walt was briefly hired as a civil engineer in maintenance at Pacific Air Command's headquarters at nearby Hickam Air Force Base prior to retiring in 1956.

The 1945 academy yearbook provides something of an insight into Wally's character. It revealed that his classmates called him "Rah Rah" and read in part, "Here is a guy who could make anyone laugh. His never-ending sense of humor, descriptions and ability to execute new pranks have kept us either amused or holding the bag … 'Rah Rah's' academy career was not effortless, but he had a way of doing things in the easiest and most effective way. We expect one of his women to snag him soon, but meanwhile his big brown eyes still have that new-fields-to-conquer look."[26]

Many years later, when Wally became famous, it was difficult to obtain an interview with his father because he generally passed the telephone or guided reporters across to his wife to answer any questions. But he did talk with pride about his son in a rare interview with Mark Waters of the *Honolulu Star-Bulletin*:

> He was full of life, extremely curious and mischievous. I never punished him; I left that to his mother. Mrs. Schirra would meet me at the door when I came home from work and tell me to go up to Walter's room, where he'd been sent, and speak to him. I would go up and say, "Mother said I should speak to you: Hello." And that would be the end of it.
>
> He was an aggressive boy but never looked for trouble. He had compassion for boys he saw going wrong and tried to help them get straightened out. He had a photographic memory. After high school he went to the Newark College of Engineering before entering the Naval Academy. His teachers in Newark told me Walter could read a book the day before an exam and pass with flying colors because he retained everything he'd read. Perhaps it's too bad it comes so easy. He breezed through the Academy while others stayed up half the night studying.[27]

In order to become eligible for flight training, Schirra was required to complete two years of "black shoe" service on board ships. As he later explained, "The sea Navy is black-shoe; the air Navy is brown-shoe – the distinction is important to us … And we are aviators, not pilots, for pilots are civilian people who bring big boats into harbors. That notion of course riles our friends in the Air Force."[28]

CHANCE ENCOUNTERS

As an academy graduate, Schirra was able to gain membership to the Army-Navy Country Club, and during one trip there his life would change forever, as these things go.

"One day I went to the club for a swim, and I noticed a willowy blonde girl sitting by the pool. She was really pretty. She was sitting alone wearing a bikini. I made up my mind to meet her. I asked a few of my pool mates who this stunning woman might be, but no one knew. I noticed that she was getting up, perhaps to leave. I couldn't lose her and I rushed over and said, 'Hi, I'm sorry, but nobody seems to know you. I'm Wally Schirra.'"[29]

Although taken aback by the young man's brashness, the lady in question stopped to talk with him. Her name, he discovered, was Josephine Cook Fraser, and she preferred to be called Jo. Schirra summoned up the courage to ask Jo to a dance at the club that evening. Although she was initially hesitant, she finally agreed. He would later find out that she had been born in Seattle and lived there until her father Donald Fraser died when she was just 13 years old. Her mother, Josephine, had eventually remarried, this time to a highly decorated naval officer, Captain William Talty Kenny. Having become a Navy Junior, Jo traveled extensively, attending a different school almost every year, including two years in Shanghai and a year in Coronado, California. After graduating, she attended Mills College in San Francisco and then joined her family on the East Coast.

That evening at the dance, Jo introduced Wally to her parents, and he was especially nervous about meeting her stepfather. "To a fresh-caught ensign wet behind the ears a captain was a god, a superior being," he reflected.[30] Sadly, Capt. Kenny would die in 1956. Two years after that Jo's mother married once again, this time to Navy Admiral James L. Holloway, Jr.

That same night of dancing at the club, Schirra introduced his Annapolis roommate, Midshipman (later Commander) John Burhans, to his sister Georgia Lou, an act of fate that would eventually lead to the two "roomies" becoming brothers-in-law.

Wally and Jo saw each other every day for the next seven days, and what began as a mutual attraction soon blossomed into love and he was determined to marry Jo after he returned from his sea duty. Meanwhile, he promised he would write to her every day.

HORSEBACK SURRENDER

Wally Schirra reported to his assigned vessel, the Seventh Fleet's armored battle cruiser USS *Alaska* (CB-1) in July 1945. The ship was stationed in Nakagusuku Bay on the southern coast of Okinawa Island (recently nicknamed "Buckner Bay" in honor of the General who commanded the U.S. Tenth Army on that island and was killed in action the previous month), which was a Seventh Fleet staging area. Schirra was serving on the ship when the war ended in August, as he later reflected:

"We saw the end of World War Two. We were in the area of Okinawa, had a few deployments. And what was really weird ... we were back in Buckner Bay, of all places, anchored, when the first nuclear bomb went. And of course everybody there said the war's over. So here we are aboard ship now and the war's over. We go up on the deck ... this large open deck, and have outdoor movies because the war's over. And all of a sudden, next to us is the battleship *Pennsylvania* [BB-38] and a kamikaze came in and dropped a torpedo on the *Pennsylvania* and almost sank it. We put out the lights, of course, cut the movies and all went back to battle stations. They hadn't surrendered ... that's the whole point."[31]

Three days after the first A-bomb was dropped on Hiroshima, a second was dropped over Nagasaki and that led the Japanese government to capitulate, ending the war in the Pacific.

As Schirra continues: "We went from Buckner Bay then, on to a tour of what they called the Yellow Sea show of force, like the old Great White Fleet. And that's how I ended up in Tsingtao, China and leading the sailors to their first liberty they'd ever had since the end of World War Two, really. It's on the Shantung Peninsula, north of Shanghai and south of Beijing (or Peking). What's so interesting about it though; here we are now, fresh-caught ensigns and we were now on the beach and the three of us, three young ensigns, rented three horses from a white Russian and went riding off in the country to see what the country was like. We were riding along and we see this stone, rugged looking building.

We look at it and we see a Japanese flag and the flag starts coming down. They were surrendering to three of us in uniform on horseback. Talk about the war not being over yet!" He recalls there being about one hundred Japanese soldiers holed up there. "Yeah, a whole fort full of them. And three little ensigns that didn't know what we were doing except riding horses. And we had to take a sword and a pistol and I think a knife from the fort leader. I don't know what rank he was because we didn't know what a Japanese officer looked like. And so we brought the sword back to the captain of the ship. We were still shaking like kids ... we didn't know what the heck happened to us."[32]

Following the Japanese surrender, *Alaska* was ordered to return to the mainland via the Panama Canal, bound for Staten Island and decommissioning. Thoughts of flying for the Navy had earlier been reinforced when a four-plane formation of F8F Bearcats buzzed the ship one day. "That was the end of my black-shoe Navy," he reminisced. Later, when his superior officer asked him if he would like to return to flight training, his immediate response was, "When can I leave?"

Wally and Jo were finally reunited, and fortunately their romance had survived the constant separations. On 23 February 1946 the happy couple wed at the U.S. Naval Academy, and left on a short honeymoon before moving into a garage apartment on Staten Island.

Ensign Schirra left the surface Navy following a temporary assignment as a briefing officer on the staff of the commander of the Seventh Fleet, Admiral Charles Maynard Cooke. Cooke's headquarters were located on the communications ship *Estes* (AGC-12) in Tsingtao, China, so Schirra flew over once again, to be followed soon thereafter by his new bride.

"We ended up ... I was on this four star admiral's staff and lived in a little beach house in Tsingtao, China. And my wife, now a fresh young bride, had three servants and said, 'This isn't bad duty. I like this.'"[33]

AN AVIATOR IN ACTION

Late in 1946, Schirra headed back to the mainland once again. "I managed to weasel my way back to the States that December and with the blessing of Rear Adm. [Fred] Boone, I was set up for flight training, initially at Grand Prairie, Texas." After 10 hours of instruction in a Stearman N2S (known colloquially as the "Yellow Peril" because of its paint scheme in the training role) he was allowed to go solo. "My wife Jo watched me as I soloed, my wings wobbling, and a few bounces thrown in." From there he transitioned to Corpus Christi, also in Texas, for training in the North American SN-J, the U.S. Navy version of the T-6 Texan. "I was designated a Naval Aviator in June of 1948 and Jo pinned on my wings." It was then that, like all other naval aviators, he was fed into what he called a "pipeline," which he explained as, "You enter as a raw naval aviator, and are spit out as a finished product."

Wally and Jo Schirra on their wedding day. (Photo: Wally Schirra personal collection, San Diego Air & Space Museum)

For his first tour he joined Navy Fighter Squadron VF-71 stationed on Rhode Island, flying the propeller-driven Grumman F8F-1 and F8F-2 Bearcats. He later described the Bearcat as his favorite type, "and always will be, even though I've hit Mach 2 in jets since."

Schirra's "favorite" airplane, a Grumman F8F-1 Bearcat flying over San Francisco. (Photo: W.T. Larkins, Warbirds Resource Group)

Schirra recalled an incident while flying in one of his beloved F8F airplanes doing some target practice. He was making an approach for landing behind a P2V, a twin-engine Neptune, an aircraft that leaves respectable traces of wake turbulence. "I had been doing inverted runs over my target, pulling just enough negative Gs to get the bullets to feed. I felt so comfortable and confident as if I had strapped the plane on. As I returned for my landing, I suddenly found myself a few feet off the ground, upside down. At the time, it seemed like no big deal to add throttle, roll around and come back for another landing." On the ground, he was immediately reprimanded by his superior. "He was standing over me while I was kissing the ground, asking me what I was doing playing in the traffic pattern. It was only afterward that I got the shakes. Those days we didn't yet know about wingtip vortices."[34]

With the onset of the Korean War in June 1950, Schirra volunteered for the Navy's exchange program with the National Guard in order to get into the action a little sooner. That same month, on 23 June, he and Jo celebrated the birth of their first child – a son they named Walter Marty Schirra, III.

Now bearing the rank of lieutenant, Schirra became a Navy/Air Force exchange pilot with the 136th Fighter Bomber Wing at Langley Air Force Base, Virginia. He was one of 25 Navy and Marine Corps pilots assigned to duty with the Air Force in exchange for 25 Air Force officers who were flying with the two other services. As he recalled:

> I reported to the 154th Fighter Bomber Squadron and with about 275 jet hours of a little over 1,000 hours total time I joined a National Guard wing of P-51 [Mustang] pilots. The wing commander had the only jet time. One hour in a T-33 [trainer]! My first flight in the Air Force was in a T-6D in February 1951. The Wing transitioned to F-84E's and we deployed to Itizuke, Japan that June.

My first combat mission was from Itizuke to Yonan, Korea and return on 23 June. As a Navy Lt. I was a flight leader and taught my team some Navy tactics which paid off during an air-to-air engagement with eight MiGs. We were attacked from six o'clock, but used the Thatch Weave to engage head on. The MiGs were surprised and pulled up and away. The next day I had to go to Seoul, Korea, to debrief the local command about our tactics.[35]

On exchange duty with the USAF in Korea, Schirra sits in the cockpit of his F-84E Thunderjet after returning from an air battle in 1951. (Photo: U.S. Navy/Wally Schirra personal collection, San Diego Air & Space Museum)

Attending a combat briefing on the Korean situation. (Photo: U.S. Navy/Wally Schirra personal collection, San Diego Air & Space Museum)

Schirra continued to see active service with his squadron in Korea, flying low-level bombing and ground-strafing runs during an eight-month tour. An outstanding pilot, he flew 90 combat missions, most of them in Republic F-84E Thunderjet fighters, and was credited with downing one Russian-built MiG-15 (NATO name: *Fagot*) fighter on 23 October 1951, as well as scoring a probable MiG-15 kill. He related the story behind his first "victory" in his book *Schirra's Space*: "On this day the MiGs had effectively tied up our F86s and were about to make it very hectic for those of us in F84s and for British pilots flying [Gloster] Meteors at our side. I spotted a MiG coming up from beneath a B29 [bomber], blazing away, and I nailed him. I did not make a classic fighter maneuver. He was slow, and I was above him – in the right place at the right time. But I was looking, and that's the art of being. It was my first MiG."[36] He was subsequently awarded the Distinguished Flying Cross and two Air Medals.

However, Schirra had one close call, "I got hit by ground fire in Korea once. The plane began to shake, and I really swabbed out that cockpit as I checked over all those instruments. Some people freeze at this point – and they die. Others just pump out all the adrenalin they have and mesh in all the gears they can and start figuring out what's wrong with this beautiful machine, and where all that complacency went. Well, in this particular plane that I was flying – it was an F-84E – I finally discovered that my right tip tank had been hit and had curled back over the aileron. This was causing it to flutter and buzz. The fluttering was feeding itself through the whole airplane until the whole works was rocking and rolling and shaking me up. I finally cut down my air speed and managed to shake the tip tank loose and drop it. The shaking stopped. The wing had a whole batch of wrinkles in it when I landed, and I knew it had been a pretty bad situation. There was not much time to think about that, though, when this airplane and I had only each other for company up in the air. At a time like this, a pilot has to be cool and collected."[37]

CHINA LAKE

After returning from Korea in December 1951, Schirra served as a test pilot from 1952 to 1954 at the Naval Ordnance Test Station at China Lake, California. He was assigned as project officer in the initial development of the heat-seeking Sidewinder missile. In the book *Wildcats to Tomcats: The Tailhook Navy*, he wrote, "My assignment was to bring a weapon from the drawing boards to fruition – and what a weapon! It was, well, right out of the space age. They called it 'Sidewinder,' an air-to-air missile that honed in on the heat of its intended victim much like its desert reptilian namesake."[38]

One of Schirra's particularly interesting memories of his time at China Lake was successfully evading a Sidewinder that had unexpectedly turned on his aircraft during one exercise.

"The Sidewinder is an extremely clever antiaircraft missile which one airplane fires at an enemy airplane to blow it out of the skies. The missile seeks out the engine of the enemy plane, just from the heat of it, and then flies right up the enemy's tail pipe before it explodes. I got to fire the first Sidewinder at a drone target to see what would happen. I was in an F3D [Douglas Skyknight] night fighter, and right after I let the Sidewinder loose it went a little haywire and started a loop which would cause it to chase me instead of the

Taking delivery of the McDonnell F3H-2 Demon fighters at China Lake. Schirra is second from right. (Photo: U.S. Navy/Wally Schirra personal collection, San Diego Air & Space Museum)

Lt. Schirra (third from left) in front of an F3H Demon. (Photo: U.S. Navy/Wally Schirra personal collection, San Diego Air & Space Museum)

Schirra engaged in test flying the F3H Demon at China Lake. (Photo: U.S. Navy/Wally Schirra personal collection, San Diego Air & Space Museum)

drone. Here was something trying to kill me, and I wasn't even mad at it. I was trying to help it along. All I could think of at the time was that I could not let this little jerk climb up *my* tail pipe. So I made a fast loop, trying to stay behind it. I simply wanted to keep its front end from ever seeing my back end. Obviously, I succeeded, although the test engineer who was with me suffered slightly from the 'clanks,' which is pilot talk for the shakes."[39]

An F3H Demon flies over China Lake. (Photo: U.S. Navy/Wally Schirra personal collection, San Diego Air & Space Museum)

There was another potentially fatal scare for Schirra early one morning at China Lake, as recalled by former U.S. Marine Edward Roy:

> In the mid-1950's at the Naval Ordnance Test Station, China Lake, California, the Navy was involved in the horizontal delivery of the atomic bomb. The Regulas system was using a ramjet engine that was not really successful for many varied reasons. The weapon designers used the same warhead (a very much improved version of the Fat Man detonated at White Sands). My involvement with the Special Weapons Project was in the assembly of the missile – the warhead [and] the propulsion system – and to document the several steps and take the moments of inertia at each interval.
>
> Wally Schirra was assigned to us as our project pilot to develop this new weapons system to the existing aircraft then in the inventory and the novel delivery method for this system. The weapons designers developed a method that was called 'the over the shoulder toss.' It was envisioned that the pilot would fly at speeds under the radar and at a prearranged point would pull his aircraft up into a climb, and at a preset angle of the attack the missile would separate [and] the rocket would fire, launching the weapon in an upward trajectory angled towards its target. The pilot would complete his maneuver rolling out of his loop and hugging the ground would speed away and be safely out of reach of the shockwave of the detonation. Wally had practiced these maneuvers at altitude, and when required to do them close to ground level he would come back and announce that these were 'idiot loops' ... from then on they were always referred to as 'Wally's Idiot Loops.'

> The incident that I refer to happened early one morning (close to 4:30 a.m.). It was important in the high desert of California (2,200 feet elevation) to get into the cooler air before the Sun was fully up and heated the air. Wally was flying a souped-up Corsair. As he got to about 200 feet, his engine coughed and sputtered and he started losing altitude… then his engine caught and started running and he started gaining altitude… then it started to sputter and cough again and he started a long slow flat turn to return to the tarmac. How many times the engine quit, I don't know, but it seemed like an eternity for him to make that long turn. At times we were sure that he was much less than 100 feet and he would crash. The fire and rescue trucks had headed out toward the south-east quadrant but he kept that aircraft in the air, and when he approached the runway he crabbed it onto the runway, dropped his landing gear, killed the engine and set it down as smooth as you please and rode the brakes to a stop about 20 to 30 yards from our exclusion area. Of course we ran to the aircraft and he had pushed back his canopy, got onto his wing and slid to the ground. I can assure you that he was pale, grim faced and said not one word! About this time his commanding officer screeched up in his Jeep. Wally got into the Jeep and sped back to the hangar. I have often said that I would have loved to have been a fly on the wall when Commander Davis got the aircraft maintenance crew together.
>
> Needless to say our project was put on hold while the aircraft was examined with a fine tooth comb. Was it fate that Wally was spared for greater things in later years as an astronaut? I think not; it was because of his superior skills and his ability to think facing stressful conditions. If he had banked into a steep return he would have cartwheeled into a fireball. He was well under the safety limit to eject and he got every inch of altitude doing what he did.
>
> I can assure you that Wally Schirra grew in stature and respect among the whole project and especially among those of us who witnessed this incident. To some it might seem as an insignificant event in the life of a test pilot; however I stood atop the torpedo crane and saw it all through binoculars.
>
> It was no mistake that the Navy selected Wally for this special weapons project and sent us their very best test pilot. It was also no mistake that he became one of our most experienced and successful astronauts, as he had the Right Stuff all along.[40]

While serving at China Lake, Schirra was presented with his Distinguished Flying Cross – the highest honor the U.S. Air Force can bestow – for his actions while flying in Korea. At a ceremony held in Los Angeles, the coveted award was presented to him by Maj. Gen. William M. Morgan, USAF. The citation accompanying the decoration read:

> Lt. Walter M. Schirra Jr., distinguished himself by extraordinary achievement while participating in aerial flight on Oct. 23, 1951. As an element leader in a flight of four F-84E-type aircraft, providing close escort on B-29-type bombers, Lt. Schirra displayed exceptional airmanship by leading his element against repeated attacks by enemy aircraft which were attacking the bomber formation in the vicinity of Namsi Airfield, North Korea. During the ensuing engagement, Lt. Schirra was credited with one possible enemy kill and one damaged enemy aircraft. As a result of the aggressiveness and ability displayed by Lt. Schirra, the B-29 bombers were able to reach their target and inflict heavy damage on the enemy airfield, despite heavy

enemy flak. By his high personal courage and devotion to duty, Lt. Schirra has brought great credit upon himself, the United States Air Force, and the United States Navy.[41]

After China Lake, Schirra served with Fleet Aircraft Service Squadron 12 as Chief Test Pilot at NAS Miramar in San Diego for the Chance Vought F7U-3 Cutlass twin-engine fighter – an aircraft that he disliked intensely, derisively calling it the "Gutless Cutlass." As he once pointed out in a memo, "I had some great experiences with the Cutlass; actually lived thru it! That airplane was a killer – the hydraulic systems were always breaking down, and if you had slats in during a stall, the airplane was not recoverable." He then received a short posting to NAS Moffett Field, south of San Francisco, with squadron VC3, which was training pilots on that same aircraft. "We were the F7U-3 Cutlass team of a transitional training unit. There were three other teams in the unit. Bob Baldwin, a graduate of the naval academy and TPS Patuxent, and today a retired vice admiral, led a team that trained on the FJ3 Fury, the Navy version of the F86. We also had teams on the F9F6 Grumman Cougar and F2H2 McDonnell Banshee. And since we cross-trained, I got to fly all of the TTU [Texas Tech University] aircraft. I no longer flew only the damn Cutlass."[42]

He also participated in Operation Steam, in which the USS *Hancock* (CVA-19) became the first aircraft carrier equipped with a steam-powered catapult capable of launching high-performance aircraft. Evaluations were carried out by launching the Cutlass in this way prior to its assignment to Fleet operations. Afterwards, Schirra returned to NAS Miramar as operations officer in VF-124 squadron, flying the all-weather McDonnell F3H-2 Demon. The last National Aircraft Show was held during the Labor Day weekend at Oklahoma City's Will Rogers Airport in 1956, with the Navy very keen to showcase its latest developments. In an unsuccessful quest for the McDonnell Trophy, Schirra flew one of four F3H-2N Demons from VF-124 on a non-stop total of 1,434 miles to the air show after launching from the carrier USS *Shangri-La*

The Chance Vought F7U-3 Cutlass; an aircraft type Schirra always despised. (Photo: Wikipedia/U.S. Navy National Museum of Naval Aviation)

(CV-38), positioned off the coast of California.[43] Also during this time, he assisted in introducing air-to-air missiles to the fleet.

In 1957 Schirra received another tour of aircraft carrier duty with VF-124, this time to the Far East and Japan aboard USS *Lexington* (CV-2). During that operational tour he became qualified as a night all-weather pilot, which would subsequently assist him with a billet at the Naval Air Test Center at Patuxent River, Maryland. Later that year, during four months at the Naval Aviation Safety School at the University of Southern California, he and Jo welcomed into the world on 29 September their second child – a daughter named Suzanne Karen.

TEST PILOT AND ASTRONAUT CANDIDATE

In January 1958 Schirra was given orders to attend the Naval Test Pilot School, Class 20, at Patuxent River in St. Mary's County, Maryland. Two of his fellow students were later astronauts Jim Lovell and Pete Conrad. Upon graduation Lovell would be named the outstanding member of that class, with Schirra and Conrad tied for second place.

Following his graduation, Schirra remained at the Naval Air Test Center as a fully-fledged test pilot with the rank of lieutenant commander. His family lived on-base in the Married Officers Quarters. He was Aviation Safety Officer in Service Test. "I was assigned as a graduate of the test pilot school, then to an organization called Service Test, which tested all the aircraft for service compatibility, and then I got to fly some of the high speed stuff and flew the very first F4H, the Phantom II. I was the 13th guy to fly it… my first

Naval Test Pilot School, Class 20, in July 1958. Schirra is at far left in third row. Jim Lovell is in the same row, third from right, while Pete Conrad is at extreme right, front row. (Photo: U.S. Navy)

Mach 2 flight was in that airplane. That was fun. Then I was flying an airplane called the F11F [Tiger], a Grumman fighter, and had some high altitude work with it and used this pressure suit... they called it a full pressure suit. What's interesting about that... that set me up for what I did later on in the space program."[44]

As he later wrote for the astronauts' book *We Seven*, "This was the most satisfying work that I could ask for as a pilot. A test pilot is fiercely proud of his profession. He knows that he is helping to develop fine new airplanes to defend his country. When you are a test pilot, you not only have to know your airplane better and fly it better than anyone else can, but you also have to be able to explain it, in complicated engineering terms, to the experts who have designed it and are building it. You have to know all there is to know about how many G forces the plane will sustain and how efficiently it burns its fuel. You have to figure out all of its idiosyncrasies and have all of its standards pinned down to a hair. Then you have to know how to apply these standards. When the plane is ready, *you* have to take the first crack at it before it has a chance to kill some other poor pilot who has to take it into battle and trust it. It's a great responsibility."[45]

In October 1958 Schirra was assigned to Edwards AFB in California, where he was involved in suitability development of the Navy's new McDonnell F4H Phantom II jet fighter. "I was the 13th to fly the first F4H, ship number one. Now I was a Mach 2 aviator in the fall of 1958."

It was during this time that he received a cryptic summons to Washington D.C. for a secret briefing at the Pentagon. He was under orders not to discuss the assignment with anyone, so even as he packed the necessary civilian clothing for his trip he and Jo were puzzled. "I reported to the Pentagon on February 2, 1959, with no inkling why," Schirra would later write. "I carried confidential orders that offered one clue – I was to dress in civilian clothes. Later I learned that President Eisenhower had insisted that the man-in-space program be a civilian endeavor. To join it, I would have to leave the Navy at least temporarily."[46]

Once inside the Pentagon, the 35 servicemen were separated out. The Navy and Marine candidates were ushered into one room for a briefing and the Air Force pilots were directed to another room. They were informed that while they would shortly be addressed by representatives from the newly formed space agency NASA (which gave them a much better understanding of why they were present) it was felt they would be far more comfortable if the initial briefing came from a senior officer within their own service. The Navy and Marine candidates were briefed by Chief of Naval Operations, Admiral Arleigh Burke, and the Air Force pilots by Air Force Chief of Staff, General Thomas D. White. Both senior officers explained that the men were there as potential space pilot candidates for the NASA program, or "astronauts" as they would come to be known. They were given an assurance that their branches of the service fully endorsed the NASA program and would similarly support anyone who chose to volunteer. They were further informed that their status would be protected, and that normal professional progress and promotions would continue.

Rear Adm. Arleigh Burke, USN, who gave the initial briefing to Navy and Marine candidates. (Photo: Wikipedia/U.S. Navy)

This, for Schirra, answered one of his lingering concerns about the impact that such a massive interruption to his career might have on any future prospects. Like many other Navy candidates in the auditorium, he felt he was on a steady, set path to eventual flag rank, and did not want to upset that goal with something as wildly unpredictable as the space program. Nevertheless, he was interested in hearing what NASA was proposing to do. "It was a tough decision to make, because I realized I was going to lose a lot of opportunities. As a naval officer, I was trained, essentially bred, to be a military aviator."

After the service briefings, all 35 candidates were directed to an auditorium for a combined NASA briefing by Charles Donlan, a senior NASA engineer and assistant to Robert Gilruth, the head of NASA's Space Task Group; Warren North, a NASA test pilot and engineer; and U.S. Navy Lt. Robert B. Voas, a psychologist who would be involved in training the future astronauts.

After outlining the NASA organization, Charles Donlan gave a brief overview of Project Mercury, the objective of which was to send an astronaut into space. Warren North then described the role of pilots in the spacecraft, and Robert Voas gave a run-down of the selection process; what would occur should they agree to continue in the screening process; and the training program that would prepare them for space flight. In conclusion, Voas said that individual interviews for those still interested would be held with members of the selection committee. This was their chance, he emphasized, to ask questions and, more importantly, to indicate whether they would continue with their candidacy or withdraw without prejudice.

"I was very reluctant to chuck thirteen years of flying to take part in what at first sounded like a stunt," was one of Schirra's more sage comments on the life-changing decision he had to make that fateful day. Meanwhile, it was impressed upon the men that the astronaut job was dangerous, and it would not be held against any man if they opted not to continue beyond that day. What really got their attention was when they were told that Project Mercury was not only a "hazardous undertaking," but of the "highest national priority."

As Schirra later observed, "I had this very vivid recollection as a child of going to the Barnum and Bailey Circus in New York and seeing this human cannonball come roaring out of a cannon. The guy wore a white suit. He came out of this cannon and landed in a net. That's how I thought of the idea of riding in a space capsule. I said, 'These guys are out of their gourds. I'm not going to do anything like that.' That was the image I had. Premature, perhaps, but I had it and said so. But people prevailed on me to stick in there for a while, to find out what it was all about."[47]

The challenge had been clearly laid out before them. Within half an hour, Schirra had accepted that even though Project Mercury might well spell the death of his naval career, he definitely wanted to be a part of it; albeit with some reservations. "That was the time to say 'No'," he recalled. "I went back to Patuxent River, where I was testing aircraft, and talked to my peer group. The same fellow who sent me to test pilot school, [Capt.] Bob Elder, said, 'Wally, if you want to go higher, farther and faster, this is the only way to do it.'"[48]

Schirra decided to press on with his application. One man who gave up the chance to become an astronaut was Schirra's brother-in-law, Navy Lt. Cdr. John Burhans who was married to his sister Georgia. Burhans, himself a 1958 graduate of the U.S. Naval Test Pilot School, now flying McDonnell F4H Phantom IIs with night-flying squadrons out of Miramar, California, declined to proceed further because he was about to enter the Naval War College at Newport, Rhode Island, on 1 July.

In a 1959 interview with *Life* magazine, Jo Schirra admitted to being "shattered" for a few moments when her husband told her what he was considering, but once he had explained everything she was excited for him.

"Some of the men who were considered for Project Mercury volunteered the first day they were told about the program. Wally came home to discuss it with me first, and I told him, 'It is a decision you have to make yourself. But please believe me when I say that if it's what you want to do, I'm all for it.' He thought about it and we talked about it for two weeks. Neither of us was concerned about the hazards of the thing, because if anything it probably will be less dangerous than test-flying an untried jet fighter. Wally thought of the decision in terms of his career. He did not want to scrap the test program he was working on to take part in a stunt. But when he knew enough about Mercury to be convinced that it was a serious program and that it was more important than the fighter he was testing, he volunteered."[49]

As Schirra later stated, "I was very much involved in a luscious project already – service testing of the F4H." He was in charge of appraising the aircraft's tactical capability and maintenance suitability. "To me, the F4H is the ultimate airplane," he said. "It is the culmination of all-weather flying. I was project officer on a plane that had reached the peak of aviation and I was supposed to dump it all."

In a later NASA interview, Schirra was asked whether he ever rued the decision. "A number of times I have regretted that, because I missed my Navy very much. I was trained to be a commanding officer, and NASA never understood what a commanding officer was."[50]

The week following the first briefings another group of potential astronauts received the same presentations at the Pentagon, with a third group scheduled for the week after that; making a total of 110 potential candidates. However, with only a nominal twelve astronaut positions on offer, and with a high volunteer rate from the first two groups, it was decided not to call the remaining group of 41 candidates to Washington.

According to Dr. Allen Gamble of NASA's selection committee, "Bob Voas and I worked late one night to tally up the results. We found that we had 32 well-qualified candidates who had passed every test, so far, with flying colors. Of the 69 who had reported, 16 had declined, 6 were found to be too tall [for the capsule], and another 15 had been eliminated by one or more of the tests. So we stopped right there and didn't call in the third group, who hadn't ranked quite as high on their records anyway. We figured that with 32 men like this we could hardly go wrong. All we had to do now was pick the very best from among these excellent candidates."[51]

One of the successful 32 candidates was Capt. Walter M. Schirra, Jr., U.S. Navy.

REFERENCES

1. Francis K. Mason, *The British Bomber Since 1914,* London, U.K., Putnam Aeronautical Books, 1994
2. Richard Stolley article, "'Flying Jennys' to Next Orbit," *Life* magazine, Vol. 53, No. 13, 26 September 1962, pp. 93–100A
3. Walter M. Schirra, Jr. with Richard N. Billings, *Schirra's Space*, Quinlan Press, Boston, MA, 1988
4. *Meriden Morning Record* (Connecticut) newspaper article, "Mrs. W. M. Schirra Receives Sad News," issue Tuesday, 11 June 1918.

5. William Olcott, *Daily Sentinel* (Rome, New York) newspaper article, "Astronaut Schirra's Mom Reminisces," Tuesday, 2 October 1962, pg. 8
6. Walter M. Schirra, Jr. with Richard N. Billings, *Schirra's Space*, Quinlan Press, Boston, MA, 1988
7. *Ibid*
8. James Cunningham, *Honolulu Advertiser* article, "Son is a Space Lindbergh, Parents Say," issue 20 April 1960
9. William Olcott, *Daily Sentinel* (Rome, New York) newspaper article, "Astronaut Schirra's Mom Reminisces," issue Tuesday, 2 October 1962, pg. 8
10. Howard Benedict, *Buffalo (New York) Courier-Express* newspaper article, "Schirra Likes Lots of Speed," issue 4 October 1962, pg. 2
11. Oradell Local History Collection, "We Have a National Hero," author unknown, courtesy of Borough Archivist George Carter
12. *Ibid*
13. *Oradell Newsletter* article "Walter M. Schirra, 1st Class Astronaut," Oradell, NJ, issue Vol. 2, Summer 2011
14. Oradell Local History Collection, "We Have a National Hero," author unknown, courtesy of Borough Archivist George Carter
15. *Ibid*
16. George Carter email messages to Colin Burgess, 24–30 June 2015
17. Richard Stolley, *Life* magazine article, "'Flying Jennys' to Next Orbit," issue 28 September 1962, pg. 96
18. *Free Lance-Star* (Fredericksburg, VA) newspaper, unaccredited article "Wally Schirra Real American Boy," issue 2 June 1962, pg. 12
19. Oradell Local History Collection, "We Have a National Hero," author unknown, courtesy of Borough Archivist George Carter
20. Walter M. Schirra, Jr. with Richard N. Billings, *Schirra's Space*, Quinlan Press, Boston, MA, 1988
21. The *Times-News* newspaper (Hendersonville, NC) unaccredited article, "Mother Recalls Schirra: Real American Boy," issue 3 October 1962, Pg. 12
22. M. Scott Carpenter, L. Gordon Cooper, Jr., John H. Glenn, Jr., Virgil I. Grissom, Walter M. Schirra, Jr., Alan B. Shepard, Jr., and Donald K. Slayton, *We Seven*, Simon and Schuster, New York, NY, 1962
23. Oradell Local History Collection, "We Have a National Hero," author unknown, courtesy of Borough Archivist George Carter
24. Jo Schirra telephone conversation with Colin Burgess, 22 February 2011
25. Wally Schirra family papers, courtesy of the San Diego Air & Space Museum Archives
26. Howard Benedict, *Buffalo (New York) Courier-Express* newspaper article, "Schirra Likes Lots of Speed," issue 4 October 1962, pg. 2
27. Mark Waters article for *Honolulu Star-Bulletin* newspaper (title and date not known). Cutting from Schirra family papers at the San Diego Air & Space Museum
28. Walter M. Schirra, Jr. with Richard N. Billings, *Schirra's Space*, Quinlan Press, Boston, MA, 1988
29. *Ibid*
30. *Ibid*

31. Fran Foley interview with Wally Schirra for The Library of Congress Veterans History Project, 19 April 2004, Rancho Santa Fe, California
32. *Ibid*
33. *Ibid*
34. Freia I. Hooper, article for unknown newspaper from Wally Schirra family papers, San Diego Air & Space Museum, issue October 1990, pg. 111
35. Wally Schirra private papers collection, San Diego Air & Space Museum
36. Walter M. Schirra, Jr. with Richard N. Billings, *Schirra's Space*, Quinlan Press, Boston, MA, 1988
37. M. Scott Carpenter, L. Gordon Cooper, Jr., John H. Glenn, Jr., Virgil I. Grissom, Walter M. Schirra, Jr., Alan B. Shepard, Jr., and Donald K. Slayton, *We Seven*, Simon and Schuster, New York, NY, 1962
38. Zeke Cormier, Wally Schirra and Phil Wood with Barrett Tillman, *Wildcats to Tomcats: The Tailhook Navy*, Phalanx Publishing Co. Ltd., St Paul, MN, 1995
39. M. Scott Carpenter, L. Gordon Cooper, Jr., John H. Glenn, Jr., Virgil I. Grissom, Walter M. Schirra, Jr., Alan B. Shepard, Jr., and Donald K. Slayton, *We Seven*, Simon and Schuster, New York, NY, 1962
40. Edward Roy, *Mountain Mail* newspaper (Socorro, NM), "Wally Schirra Was a Great American," issue 17 May 2007, pg. 8
41. Unaccredited newspaper clipping from the Schirra family collection, San Diego Air & Space Museum
42. Zeke Cormier, Wally Schirra and Phil Wood with Barrett Tillman, *Wildcats to Tomcats: The Tailhook Navy*, Phalanx Publishing Co. Ltd., St Paul, MN, 1995
43. McDonnell *Airscoop* magazine, unaccredited article, "One of the Astronauts is Old Friend of M.A.C. Family," issue June 1959
44. Fran Foley interview with Wally Schirra for The Library of Congress Veterans History Project, 19 April 2004, Rancho Santa Fe, California
45. M. Scott Carpenter, L. Gordon Cooper, Jr., John H. Glenn, Jr., Virgil I. Grissom, Walter M. Schirra, Jr., Alan B. Shepard, Jr., and Donald K. Slayton, *We Seven*, Simon and Schuster, New York, NY, 1962
46. Untitled typed paper by Wally Schirra, from the Schirra family papers held at the San Diego Air & Space Museum
47. Zeke Cormier, Wally Schirra and Phil Wood with Barrett Tillman, *Wildcats to Tomcats: The Tailhook Navy*, Phalanx Publishing Co. Ltd., St Paul, MN, 1995
48. John Zarrella, CNN, video interview with Wally Schirra, aired in tribute 3 May 2007
49. Jo Schirra, article, "Maybe I've Been Lucky," *Life* magazine, issue 21 September 1959
50. Walter M. Schirra, Jr., interviewed by Roy Neal for NASA JSC Oral History program, San Diego, CA, 1 December 1998
51. Allen O. Gamble, *Personal Recollections of the Selection of the First Seven Astronauts*. Paper presented at the Men's Club of the Bethesda United Methodist Church, Bethesda, Maryland, 10 March 1971. Courtesy of Kris Stoever

2

The making of a flight-ready astronaut

Wally Schirra almost didn't make it. Having shed a few excess pounds during tough and invasive physical and psychological testing at the Lovelace Clinic in Albuquerque, New Mexico, the next step in the Mercury astronaut selection process involved stress testing together with his fellow 31 candidates at the Wright-Patterson Aero Medical Laboratory in Dayton, Ohio. For a brief time, and much to his consternation, Schirra thought he had been scrubbed out of contention because of a previously undiagnosed medical condition.

During his medical examination at the Lovelace Clinic, a specialist examined Schirra's throat and seemed to be paying particular attention to his larynx. The fact that he was a heavy smoker became an immediate cause for concern to Schirra, especially when the doctor called in a colleague who also carefully scrutinized his throat. They went away to discuss their findings, and shortly after returned with the facility's founder Dr. Randy Lovelace in tow. Lovelace was also chairman of NASA's Special Advisory Committee on Life Science and therefore a key player in the selection of the Mercury astronauts. He delivered Schirra some disturbing news.

"I had a polyp, a chunk of swollen membrane, on my larynx. One of the doctors mentioned the word 'tumor,' and then I really started to sweat. But everyone assured me it was benign, and they offered to take it out right there. I suggested we ought to consult the Navy first… Dr. Lovelace said, 'Don't worry. There's no problem. You can take care of this when you go back to Patuxent.' The fact that they were willing to work on me themselves, though, was the first glimmer I'd had that maybe I was going to make the team and was being separated from the other boys."

Schirra had his operation performed as a priority and the polyp was removed at the Bethesda Naval Hospital on 3 April 1959. He was instructed not to speak for the next four days. "That, of course, was the bona fide proof that I was on the bandwagon and that they really cared."[1]

Although it was to help propel him into the most exciting undertaking of his life, Schirra was never a big fan of the probing tests that he and the other finalists had to endure at the Lovelace Clinic, as he later expressed in his autobiography.

C. Burgess, *Sigma 7*, Springer Praxis Books, DOI 10.1007/978-3-319-27983-1_2

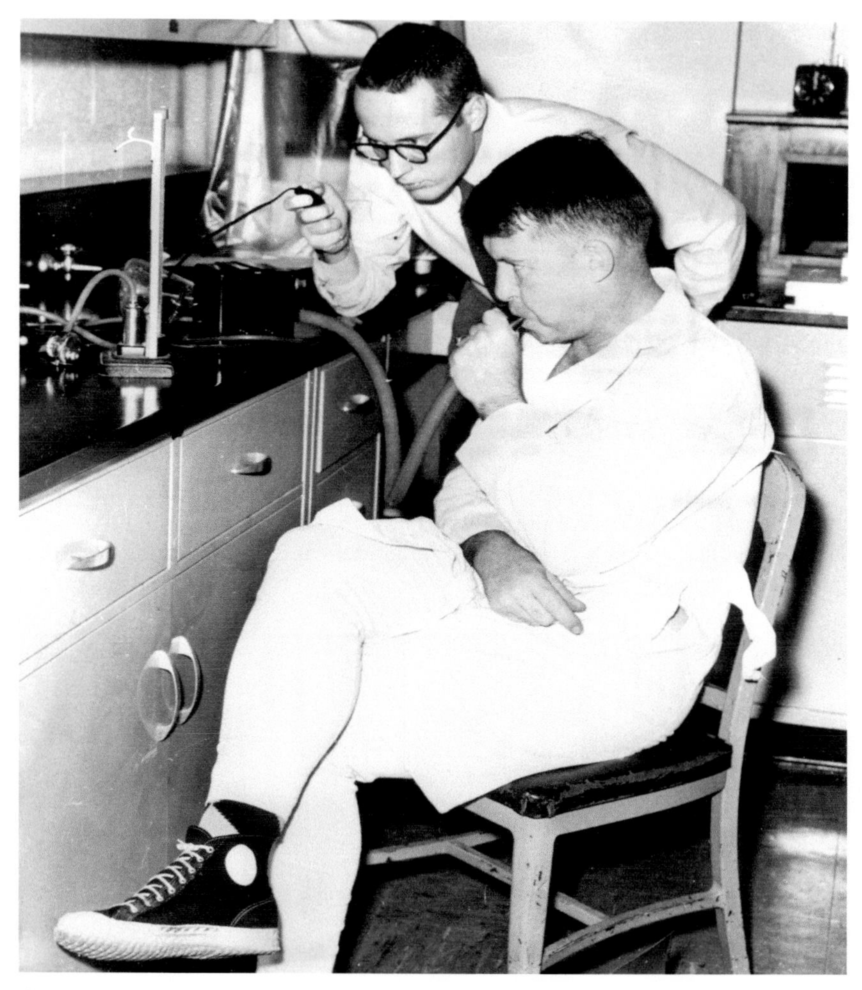

Wally Schirra undergoes a lung capacity test at Wright-Patterson Air Force Base. The goal for each astronaut candidate was to maintain pressure as long as possible by blowing into the mouthpiece to support a column of mercury higher than 40 mm for as long as possible. Once the column dropped below that level the clock was stopped. (Photo: USAF)

"I still feel that the physical exams at Lovelace were an embarrassment, a degrading experience. I have said many times – and meant it – that it was a case of sick doctors working on well patients. I make the point in talks to medical associations. It was a rare, almost unheard of situation in which so many healthy individuals submitted to an array of tortures – proctoscopies, barium-enhanced X-rays, psychological interrogation and so on."

Schirra understood completely that the doctors at the Lovelace Clinic were simply trying to establish a physiological and psychological guideline for use in tests on later space flights, and admitted that – at the time – it was felt to be mostly a valid exercise. However he would later discover through NASA flight surgeon Bill Douglas, a close friend, that none of the punch-card computerized data obtained at the Lovelace Clinic could be interpreted, and all of the tests had to be redone after those who were selected had joined the program. "Much as I'd rather not knock a dead man," he later wrote, "Randy Lovelace did a lousy job."[2]

On the first day that he was permitted to speak following his operation, 7 April, Schirra received a phone call from NASA's Charles Donlan telling him he had been selected, and asking if he was still interested in joining the space agency. If so, he had to arrange to be in Washington two days later for the formal announcement at a press conference. He had indeed "made the team" as one of the seven Mercury astronauts. Suddenly, as Schirra later described, he was faced with "the most momentous decision of my life. This was my last chance to say, 'Hell no, I'm not going to chuck my Navy career.'" But he had already given the question a lot of thought; he set any misgivings aside and decided to accept the challenge.*

ON BECOMING AN ASTRONAUT

At 2:00 p.m. on Thursday, 9 April 1959, the nation's first astronaut group was introduced to the world by NASA Administrator T. Keith Glennan during a crowded press conference held at the agency's temporary headquarters at the Dolley Madison House in Washington, D.C.

The Mercury astronauts are introduced to the world's media. Left to right: Deke Slayton, Alan Shepard, Wally Schirra, Gus Grissom, John Glenn, Gordon Cooper and Scott Carpenter. (Photo: NASA)

* For a full description of the selection process, see the author's 2011 Springer-Praxis book *Selecting the Mercury Seven: The Search for America's First Astronauts.*

> Ladies and gentlemen. Today, we are introducing to you and to the world these seven men who have been selected to begin training for orbital space flight. These men – the nation's Mercury astronauts – are here after a long and perhaps unprecedented series of evaluations which told our medical consultants and scientists of their superb adaptability to their upcoming flight. Which of these men will be first to orbit the Earth, I cannot tell you. He won't know himself until the day of the flight.
>
> The astronaut training program will last probably two years. During this time our urgent goal is to subject these gentlemen to every stress, each unusual environment they will experience in that flight. Before the first flight we will have developed the Mercury spaceship to the point where it will be as reliable as man can devise. We expect it to be as reliable as any experimental aircraft.
>
> It is my pleasure to introduce to you – and I consider it a very real honor, gentlemen – from your right, Malcolm S. Carpenter, Leroy G. Cooper [he missed giving the Jr. suffix], John H. Glenn, Jr., Virgil I. Grissom, Walter M. Schirra, Jr., Alan B. Shepard, Jr., and Donald K. Slayton ... the nation's Mercury astronauts![3]

If the space agency had been expecting an air of respectful decorum in making the announcement, it was in for a shock. Press cameramen surged toward the front of the stage, shouting out to the seven men, eager to get the best possible photographs. The overwhelming reaction to the announcement stunned everyone, not least the newly-named astronauts.

"We were seven veteran test pilots but unsophisticated young men in many ways, not very well prepared for the sudden fame of being America's first astronauts," Schirra would later write in his memoir *Schirra's Space.* "We went to Washington in April 1959 for a press conference which was a scary event, as we faced a thundering herd of reporters and photographers."[4]

One vexing problem Schirra had in his early days as an astronaut was the frequent mispronunciation of his Swiss-heritage surname. Whenever someone got jammed up on it, he would patiently explain to them that it rhymed with "Hurrah." One day there was a televised ceremony involving the seven Mercury astronauts, and an obviously unrehearsed reporter was going along the line with a piece of paper on which was written their names, introducing them to the television audience. When he got to Schirra he hesitated, squinted a little, and mumbled "Walter Marty ... um ... Sky ... ray!" Everyone except Schirra, who was more than a little peeved, got a huge laugh out of it later on, especially Alan Shepard, who would thereafter seek any opportunity to jokingly call him "Captain Skyray." The F4D Skyray was actually the name of a Navy aircraft type which Schirra had flown at Patuxent River, so in a way it was rather an appropriate, if inadvertent, nickname. And it wasn't only Shepard who used to tease Schirra in that way – even his wife Jo would often call him "Skyray." But as he later revealed he actually came to like the nickname, and even used it as his CB radio call sign.

The Schirra family at home following the announcement of his selection as a Mercury astronaut. (Photo: NASA/Life)

EARLY ASTRONAUT TRAINING

From the outset, there were few, if any, facilities available to support the training of the Mercury astronauts, as outlined by Robert B. Voas of NASA's Space Task Group in *The Training of Astronauts: Report of a Working Group Conference*, prepared by The Armed

A Douglas F4D-1 Skyray from the Navy Test Pilot School, Patuxent River. (Photo: Greg Phelps/Airliners.net)

Forces – NRC (National Research Council) Committee on Bio-Astronautics in 1961. The report disclosed that training devices and manuals only began to appear in the first 12 to 15 months following the selection of the astronauts. In fact, the more elaborate and complete training devices were not in operation until more than a year later. Consequently, the early portion of the astronaut training program depended on reviewing design drawings of spacecraft components and on travel to various Mercury production plants and other facilities to attend design briefings. There was a strong reliance on presentations by scientists and engineers from the Space Task Group and representatives from the prime contractors.

The existing Armed Forces Aeromedical facilities were another valuable resource because they would familiarize the astronauts with many of the conditions of space flight. As the report stated:

> About one-half of the program which has resulted from these considerations is allotted to group activities, and the other half to individually planned activities in each astronaut's area of specialization. A review of the astronauts' travel records reveals the relative division of their time between group training and other duties associated with the development of the Mercury vehicle. During the six-month period from July 1 to December 31, 1959, the astronauts were on travel status almost two months, or one out of every three days. Half of this travel time (28 days) was

Wally Schirra at his desk in 1959 with Alan Shepard in the background. In the bottom photo they are joined by Gordon Cooper. (Photos: Wally Schirra personal papers, San Diego Air & Space Museum)

spent on four group-training activities: a centrifuge program; a trip to [the] Air Force Flight Test Center, Air Force Ballistic Missile Division, and Convair; a weightless flying program; and trips to fly high-performance aircraft during a period when the local field was closed. The other half of their travel time (27 days) was devoted to individual trips to attend project-coordination meetings at McDonnell and the Atlantic Missile Range, or for pressure-suit fittings, couch moldings, and viewing of qualification tests at McDonnell, B. F. Goodrich Company, and their subcontractor plants. These individual activities, while providing important training benefits, are primarily dictated by the Project Mercury development program requirements and are not considered part of the group training program.[5]

In this photo taken 25 September 1959 in San Diego, California, six astronauts stand around models of the Atlas and Mercury capsule designed to carry them into space. From left are Wally Schirra, J. R. Dempsey (executive of Convair, maker of the Atlas rocket), John Glenn, Gus Grissom, Alan Shepard, Gordon Cooper and Scott Carpenter. Deke Slayton was not present. (Photo: NASA/Convair)

There was an unsustainably large volume of information the new astronauts had to study and assimilate; it was a learning process an astronaut once likened to "drinking from a fire hose." While they were all involved to an extent in spacecraft and booster familiarization, a way was needed to lessen the burden individually. In recognizing this, each astronaut was handed an assignment to focus on a particular area of specialty that best reflected their aviation and engineering background. Scott Carpenter took on communications and navigation because of his experience in celestial navigation and airborne electronics. Gordon Cooper's tasking was the development of the Redstone booster, and Deke Slayton had a similar assignment, but with the far more powerful Atlas rocket. Meanwhile, John Glenn followed through on his particular interest by working on the interior design and instrument layout for the Mercury capsule. Gus Grissom dealt with its control systems. Alan Shepard took on tracking and recovery operations. The Mercury capsule's life support systems and spacesuit design became Schirra's responsibility. It was a task he found both interesting and enjoyable.

Schirra's involvement with the Mercury space suit came quite early in their program schedule. In July 1959, following an extensive evaluation process, NASA handed the

B. F. Goodrich Company in Akron, Ohio a contract to develop a suitable pressure suit that could safely and comfortably be worn by the early astronauts. The suit they came up with was based on the Mk. IV full pressure suit that the company had designed and manufactured for the U.S. Navy.

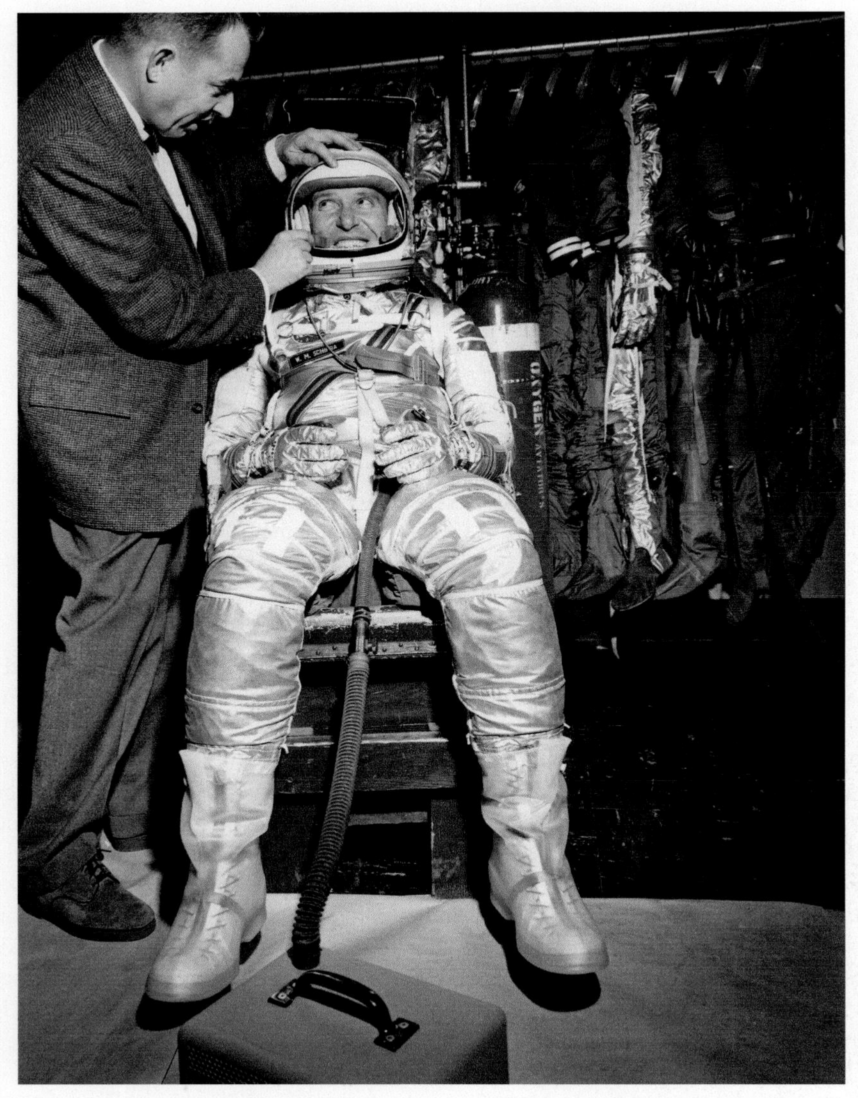

The design and integrity of the Mercury space suit was a responsibility that Schirra took on. (Photo: NASA)

"I acted as a kind of consultant on the pressure suit," he later commented. "It had a number of things wrong with it when we first started the program. But fortunately we had time to play with it – in fact, we nit-picked it to pieces – and we wound up with a very useful gadget."

He described the pressure suit as "the skintight cocoon of rubber and fabric into which we seal ourselves before each flight or practice mission. Essentially, the suit is a man-shaped balloon. It is also tailor-made, and it fits us so snugly that we have to use thirteen different zippers and three rings to put it on. Complete with helmet, the suit weighs about 22 pounds, and each one costs about five thousand dollars [in early 1960s terms]."[6]

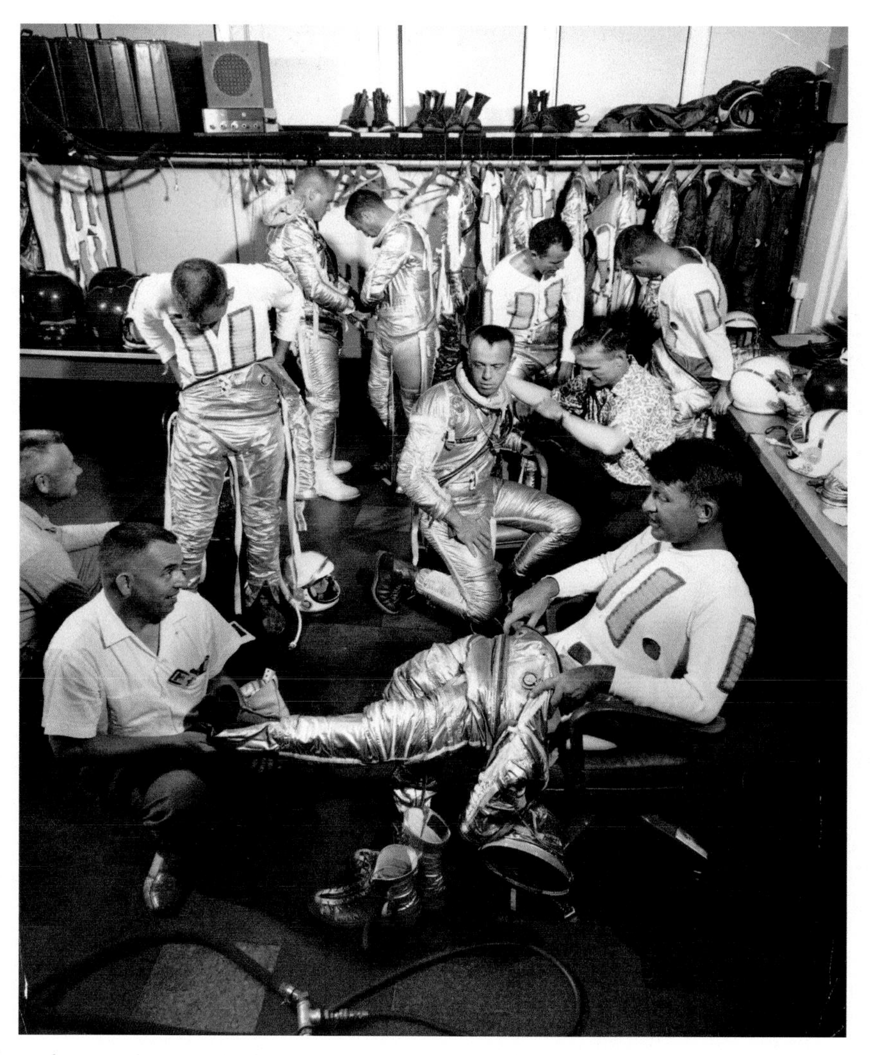

A rarely seen photograph of all seven astronauts being fitted into their space suits in preparation for a photo shoot in 1960. As Schirra emphasizes, it was the only time that all seven men were suited up at the same time. (Photo: NASA)

This publicity photograph of the Mercury astronauts was taken later that day. (Photo: NASA)

In discussing his overall assignment to the Life Support System with interviewer Shirley Thomas in 1961, Schirra described the Mercury capsule as "a little bit of Earth" transplanted into the heavens:

> What the Life Support System really amounts to is a maintenance of a livable atmosphere, as closely approximating Earth as we can manage. Ideally, this turns out to be about one-third of the atmospheric pressure at sea level, with a 100 percent oxygen environment. This will then give us the sea-level oxygen supply that we have lived under all our lives. This is done by storing oxygen in bottles at very high pressure – 7,500 pounds per square inch.

> It is metered through a system and is fed to the astronaut through the capsule environment, or through the pressure suit that he wears. The astronaut can either live off the pressure suit or live off the cabin itself. The oxygen that is supplied to the man must be cleaned – the carbon dioxide is removed by a lithium hydroxide chemical separation, the water vapor is removed through a sponge, and then is collected in a concentrate tank. Odors and solids are caught in an activated carbon filter and in a solids filter, respectively. Another thing we need, in order to remain alive, is water. We have drinking water with us.
>
> [The capsule] is basically a thermos bottle in space. We have to get rid of the heat of the man and the heat of the electrical equipment that is working. This is done through a water-cooler system. In space, water boils at a very low temperature, so by heat exchange we cool off the oxygen that is running through the cabin, and through the pressure suit. We will maintain a normal room temperature; even on re-entry, the pressure suit keeps it well within tolerable limits.[7]

One by one, the seven astronauts traveled to Akron to be individually fitted for their pressure suits. After he had stripped down to his long johns, the Goodrich technicians would plaster him all over with strips of paste-impregnated paper. This would dry into a stiff mold which they would carefully cut off the astronaut and later use to make each suit – a process that took about a month to complete. Refinements would later be made in consultation with the astronaut until everyone was satisfied.

After having their individual suits created at Goodrich, the astronauts next made their way to a subcontractor for helmet fittings, and then traveled on to the Navy's Air Crew Equipment Laboratory in Philadelphia, Pennsylvania for tests while wearing their space suit under extreme conditions.

The integrity of the suits was a crucial element of future flights, and therefore they had to be subjected to a range of wear and tear tests to ensure they would stand up to the task. One of the tests Schirra endured on behalf of the astronauts was a highly uncomfortable few minutes seated inside a sealed heat chamber, where the temperature was ramped up to 180°F to verify that the suit and its oxygen system could withstand extreme heat. "To make absolutely sure, we wanted to have an astronaut sealed up inside the suit so he could see for himself," he revealed. "I made this heat run. There was no need for all of the others to do it. All they needed to know was how the test went and what I thought about it." Meanwhile, his fellows were busy investigating other aspects of the Mercury program and they would all trade information with each other at briefing sessions. "All of us were vitally *interested* in everything, of course. But we did not have the time to *master* everything."[8]

WALLY AND THE TURTLE CLUB

One word that will always be associated with Wally Schirra is the slang term "gotcha," which means exactly what it says. While all of the astronauts participated in playing pranks on each other to alleviate the mental strain of training and tuition, Schirra was the undoubted master. One that he recalled in his autobiography was a trick they all played on an unsuspecting Dr. Stan White, who was the chief flight surgeon for the space program.

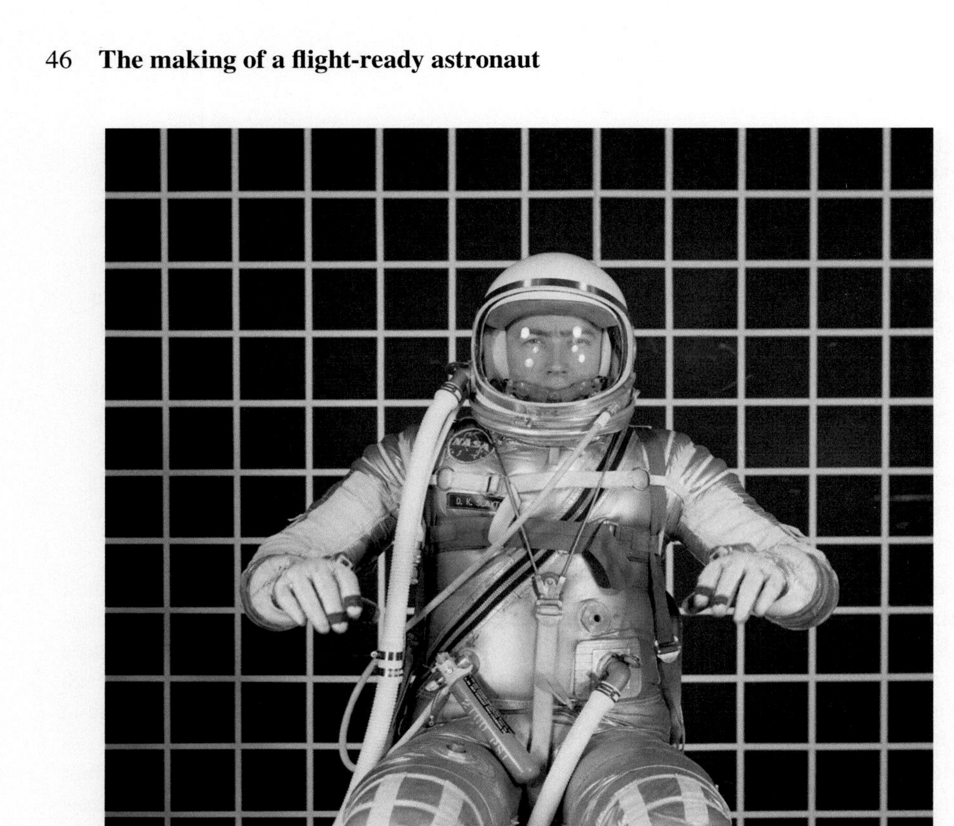

A suit technician wearing Deke Slayton's fully pressurized Mk. IV space suit during a mobility demonstration. (Photo: B. F. Goodrich)

"White drove a new sports car, and he liked to brag about its efficient performance," Schirra recalled with glee. "So we plotted his comeuppance. For a week we added gasoline to his tank, a pint a day, and he raved about the great mileage he was getting. The following week we siphoned off a pint a day, and he went berserk. White never did figure it out."[9]

In this 1955 photograph, X-15 pilot Scott Crossfield undergoes a similar extreme heat test of an XMC-2 pressure suit. (Photo: NASA)

The astronauts also agreed that due to their fitness regime they should quit smoking, with the exception of Gordon Cooper, who was already a non-smoker. Schirra himself was a pack-a-day man, but he said this abstinence did not last long for any of them. He would become tense and irritable, and when he did, Jo would threaten to kick him out of the house. But even when they smoked they found childish amusement in lining the metal ashtrays on their desk with a thin film of gasoline. Whenever someone took a seat opposite them, they would drop a little hot ash from their cigarette into the ashtray, which erupted in a sudden flash of flame. "Fiendish but fun," he recalled.[10]

In his pre-Corvette days, Wally Schirra shows his Austin Healey roadster to NASA flight surgeon Bill Douglas. (Photo: NASA)

John Boynton was an engineer working on the Mercury program. In 2009 he told a NASA interviewer about another long-running joke that gave Schirra a lot of fun over the years.

> Since I'm telling secret stories, this is a good time to talk about Wally Schirra. Wally was a very colorful guy. I liked Wally a lot, and whenever he was at the Cape preparing for the flight, if I was down there, we would get together and have breakfast. The neat thing about Wally – he was driving an Austin Healey 3000 roadster, which is a classic British sports car, and I was driving one too, so we had a common bond besides airplanes. As you might know, nearly all of the original Mercury astronauts eventually drove Corvettes, but that was after Schirra's flight, otherwise he would have only talked about his hot Corvette. So he flew MA-8, and we used to talk about that, both before and after the flight.
>
> Anyway, having breakfast with him one day, he said, "John, you want to join the Turtle Club?" I said, "What's the Turtle Club?" He said, "Well, it's kind of a joke. A bunch of us guys got well-oiled one night, and we decided we were going to start a new brotherhood called the Turtle Club. The gimmick is that you should know what to say when someone asks you the Turtle-Club question." I said, "What question?" He grinned broadly and said, "You ask someone, 'Are you a turtle today?' And if he's a member of the Turtle Club, he'll tell you what you're supposed to say, but if he's not, he'll say, 'No, I'm not a member of that club, why?' He's then informed that, according to an old tradition, he has to buy everyone a round of drinks. Of course, the guy can say, 'No, I refuse to buy you losers any drinks,' and walk out. But that's the deal. You must buy a round of drinks if you don't say the correct thing."

> I said, "Well, what's the correct thing?" He says, still grinning, "'You bet your sweet ass I am.'" Then with a straight face, he said, "John, are you a turtle today?" I immediately shot back, "You bet your sweet ass I am." So he made out a funky little green card, reproduced on an office Xerox machine, and I became one of the original members of the Turtle Club. It was a joke with most of the original seven Mercury astronauts and their drinking buddies, except probably John Glenn. Glenn never drank and never partied; he didn't go out to the bars. But the other five guys would have joined. Even today, 45 years later, in the NASA area right on the lake, there is a joint called The Turtle Club, in honor of Wally and his drinking buddies.[11]

Wally later refined the rules so that anyone wishing to join the Turtle Club first had to answer one of four questions, each of which required what seemed like a vulgar response. One example: What does a woman do sitting down that a dog does on three legs, and a man does standing up? The obvious answer would be a derivative of "urinate," but the correct answer to this particular question was "shake hands," as conventional etiquette required that a man needs to stand up in order to shake hands, while a lady can remain seated.

MA-8 PILOT

In mid-March 1962, Wally Schirra was one unhappy astronaut. The pilot originally named to the MA-7 flight, Deke Slayton, had been diagnosed with atrial fibrillation, a slight and intermittent heart problem. As a result, a devastated Slayton was taken off flight status and removed from the second orbital Mercury mission. As his backup pilot, the flight should have been automatically handed to Schirra, but this did not happen, much to his consternation.

Acting on the advice of Walt Williams, then head of NASA's Space Task Group, Manned Spacecraft Center Director Robert Gilruth decided to name Carpenter as the replacement MA-7 pilot, with Schirra reassigned as his backup. Williams had figured that as the three-orbit mission was likely to be little more than a repeat of John Glenn's flight the previous month, it would be best to give it to Scott Carpenter. He had served as Glenn's backup for more than five months, and logged more training hours for an orbital mission than either Slayton or Schirra. Time was running short to get MA-7 up and running, and Williams knew that NASA was contemplating a seven-orbit flight later in the year, which he felt would be better suited to Schirra. But Schirra was far from pleased with losing the flight, firmly believing it should have been his.

Carpenter ended the troubled MA-7 on 24 May 1962 by landing 200 miles beyond the intended splashdown due to a delay in firing the retrorockets and an off-nominal attitude. This meant it would take a while to recover the astronaut and his spacecraft. Schirra was annoyed at the media coverage which spoke of a "lost" astronaut and the possibility that Carpenter may have died during re-entry.

"We knew where Scott was," Schirra later told a *Newsweek* reporter. "We had many good radar fixes on Scott. I think the people who were relaying information out of Mercury Control Center boo-booed in not saying that we knew where Scott was … We had telemetry after the [communications] blackout which meant he had gone through the G-pulse … yet

Wally Schirra in discussion with Deke Slayton. (Photo: NASA)

it never came out and it was disgusting. I don't feel badly about harpooning someone on this, because if they do it to me, I'm going to be furious."[12]

He was certainly outspoken, but that was just his nature. As expected, Schirra was assigned to the next mission. The official announcement came on 27 June 1962. The MA-8 mission, then scheduled for late September, would be flown by Wally Schirra, with Gordon Cooper as his backup. The news bulletin further reported that the flight would last "for as many as six orbits."

As Schirra commented following the announcement of his selection for MA-8, "I am very, very thrilled and looking forward to doing a job which is important for all of us."

In his autobiography Schirra said that he had "lucked out" because Carpenter's flight was very much just a rerun of Glenn's. Although he felt quite strongly that he had been deprived of what he felt was his turn, he was eventually pleased it had happened that way. "I said I wanted six orbits. [Robert] Gilruth and [Walt] Williams refused to make a commitment. They were worried about the fuel supply. John and Scott had used up the hydrogen peroxide that powers the attitude control thrusters. My mission would be open-ended, but a recovery force would be deployed in the Pacific Ocean. If I did go six orbits, I would land about 275 miles northeast of Midway Island."

The earlier Mercury missions had been flown in what Schirra described as "chimp mode" (a derogatory term for flying on automatic mode), which he said unnecessarily

Scott Carpenter on board the USS *Intrepid* following his MA-7 flight. (Photo: NASA)

wasted precious fuel. He argued that he could control the spacecraft's attitude by hand, shutting the spacecraft down once it was properly aligned in orbit and simply drifting. He would return to retro control using the manual controls well before the retrorockets were to be fired for re-entry, and so save a significant percentage of fuel. "I wanted to show how well I could operate the control system. I would shut down and start up and return to retro attitude by sighting on the horizon and the stars. We trained for that at a planetarium at the University of North Carolina in Chapel Hill."[13]

Wally Schirra and Scott Carpenter studying star charts with Dr. Jocelyn Gill (NASA's Chief of In-Flight Sciences) and an unnamed planetary scientist at the Morehead Planetarium in North Carolina. (Photo: NASA/Retro Space Images)

The decision on the number of orbits Schirra would fly depended on a number of technical factors, NASA explained, and these would be evaluated constantly up to the time of flight and during the first orbits of the Earth. If the mission went the full six orbits it would involve a nine-hour flight, which was double that of the two previous Mercury missions.

Additionally, a flight of five or six orbits would mean landing in the Pacific Ocean, whereas the first four manned Mercury flights – two suborbital and two orbital – had splashed down in the Atlantic. Landing points for one, two and three orbits on MA-8 would remain the same as the earlier Mercury-Atlas flights, off the south-east coast of the United States. If the flight only completed four orbits the spacecraft would splash down 200 miles east of Midway Island.[14] As was outlined by Wes Oleszewski of *Aero News Network*:

> *Sigma 7* was the first Mercury flight to be extended beyond the Project Mercury's original design planning of three orbits. This brought into play several problems, the greatest of which was the Earth rotating on its axis. During a three-orbit flight, such as that of John Glenn and Scott Carpenter, the Earth rotated about 67.5 degrees underneath the plane into which the spacecraft had been inserted. Thus, NASA had placed all of its ground tracking and communication assets within range for such flights. On Schirra's six orbit flight, however, the Earth would rotate a total of 135 degrees under *Sigma 7*. That meant that Schirra's spacecraft would work its way farther and farther south and would be out of range of more and more of the established stations on orbits four, five and six. Additionally, a retro-fire in orbit six would splash *Sigma 7* down in the Pacific rather than the Atlantic.[15]

Giving his insider's take on what lay ahead for Schirra and the flight of *Sigma 7* was Manfred ("Dutch") von Ehrenfried, at that time a flight controller in NASA's Flight Control Operations Branch. He commented that planning for the MA-8 mission began right after the successful completion of John Glenn's orbital mission. "It was to be an intermediate step to an extended duration flight which at the time was planned for 18 orbits. Extensive work was done on the spacecraft in order to extend its power and life support capabilities. A lot of analysis of MA-6 and MA-7 gave the engineers the data that they needed to make the desired changes. A look at the recovery planning for extended duration flights led to the conclusion that a mission rule for contingency recovery of the astronaut would be violated. He required to be recovered within 18 hours after landing. As a six orbit mission would meet that requirement, the original plan for seven orbits was reduced to six."

As von Ehrenfried also explained, Schirra's mission was intended to be more focused on engineering rather than science. "Only four non-engineering scientific experiments were planned, two of which were completely passive, so his active involvement was required in only two of them." Furthermore, he was tasked with an evaluation of the problems Scott Carpenter had faced in determining his yaw, which had contributed to his overshooting the planned landing zone. On MA-8, "Schirra was to use both the periscope and the window in daylight and in darkness, and conserve his fuel by minimizing his use of the RCS [Reaction Control System] thrusters. He practiced these maneuvers in Mercury Procedures Trainers at Langley and at the Cape."[16]

In a 1972 interview with Jerry Bledsoe of *Esquire* magazine, Schirra spoke about how all of the Mercury astronauts loved to fly, but took their flying seriously. They were accustomed to taking risks, he pointed out, but not inordinate risks, and had taught themselves to function well under pressure. Fear was not allowed. Apprehension was their word. "I felt we had a lot of apprehension, but I didn't believe that we were, in effect, for example, putting on our scarf and goggles. You know, launch me, baby! It wasn't that. We were engineering test pilots, very actively involved. We had a long time to face up to what we were doing, three years on the average, and if there was something we had any fear of, we would train it down to apprehension so we wouldn't have any surprises, so we could perform. We literally trained out fear."[17]

The month following his appointment to MA-8, Schirra was able to give more details on his flight during a televised broadcast from Hangar S at the Cape that was beamed to most countries of the world through the Telstar communications satellite.

Schirra practicing water egress techniques off Cape Canaveral. (Photos: NASA)

After another water drill, a drenched Schirra is "helped" by John Glenn and an instructor. (Photos: NASA)

Training included practicing simple tasks while flying parabolic arcs in jet aircraft to create short spells of weightlessness. (Photo: NASA)

"My flight plan calls for up to six orbits ... nine hours in space," he stated. "There are only a few equipment changes. We were able to take out a radio transmitter and receiver by making the main command radio do double duty and work on Earth as well as in space. We added a new antenna to make it work better. I also have a new small radio to use from a life raft if I decide to leave the spacecraft while waiting to be picked up. There are a couple of control circuit changes; parachute wiring for one, and a new switch to insure against using too much fuel during flight for another."[18]

By late July, NASA was quietly confident that it would be able to launch the MA-8 mission on 18 September. The spacecraft to be used, No. 16 from the McDonnell plant, had been modified and checked out to allow for a flight of up to seven orbits. But the Atlas 113-D booster that was to be used did not arrive at the Cape until 8 August, and then the Air Force required a static firing of the propulsion system because the vehicle incorporated a baffle injector modification to eliminate the standard two-second hold-down before the rocket was unleashed from the launch pad. This static test, even if it was a success, would require postponing the launch until 25 September.

According to the flight program which Schirra had painstakingly worked out with Richard ("Dick") Day, Assistant Chief of the Manned Spacecraft Center's Flight Crew Operations Division, during his first two orbits he was to establish in detail how all of the spacecraft's systems were functioning, then he would fly at least one full orbit on the automatic stabilization control system (ASCS) that had proven so troublesome on previous, shorter flights.

The MA-8 flight plan called for more drifting flight than on the two previous orbital missions. The supply of hydrogen peroxide fuel, and its rate of usage, had come to be regarded as the most critical items in accomplishing the mission; particularly after the fuel consumption issue that had befallen Scott Carpenter on his MA-7 flight. The rate of fuel usage would be largely determined by the mode that Schirra used in controlling *Sigma 7*, and previous experience had demonstrated that the ASCS mode was the most economical of the four control modes, and the manual mode the most expensive.[19]

There had been a lot of fall-out from Scott Carpenter's flight, much of it to do with the fact that his five hours in space had been overloaded with a jumbled array of tests and experiments. This mistake would not be repeated on Schirra's flight, as explained by historian David Baker in *The History of Manned Space Flight*:

> The lessons from Scott Carpenter's MA-7 flight were clear and unmistakably distinct. The flight plan had been excessively loaded with a myriad activities, each the product of a scientist or an experimenter anxious to have the astronaut obtain the best possible data for his project. The real trouble was that flight plan coordination had not really been applied, and that mistake was more a product of poor integration than any deliberate disregard for the pilot's need to allocate time for systems management – housekeeping, as it became known. One particular experiment would have the astronaut position his capsule in a certain attitude, the next task would call for the spacecraft to be in a totally different attitude, while a third activity would have Mercury return to angles only marginally different than those for the first operation. The need to place the first and third activities in a time sequence, followed by the second after that, was only brought home in the post-flight de-briefing sessions.

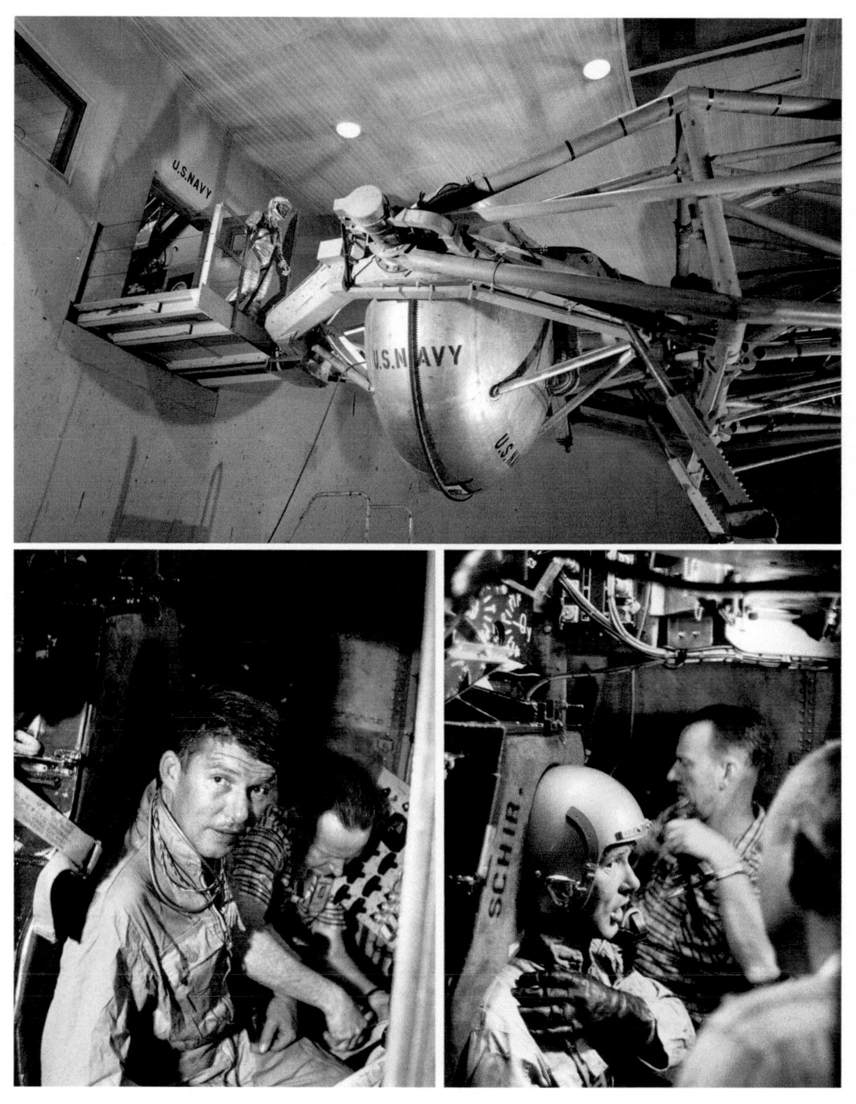

Although he loathed it as a training device, believing it to be unnecessary for a qualified test pilot, Schirra took part in g-force exercises in the centrifuge at the Naval Air Development Center in Johnsville, Pennsylvania. In the top photo he is about to enter the gondola, and in the bottom images he is being prepared for a centrifuge run. (Photos: NASA)

Intelligent management of the spacecraft's attitude, where experiments would be placed in an attitude-dependent sequence so that the capsule could sweep across successive pitch or roll arcs rather than move continuously back and forth, became a prime consideration for all future missions.[20]

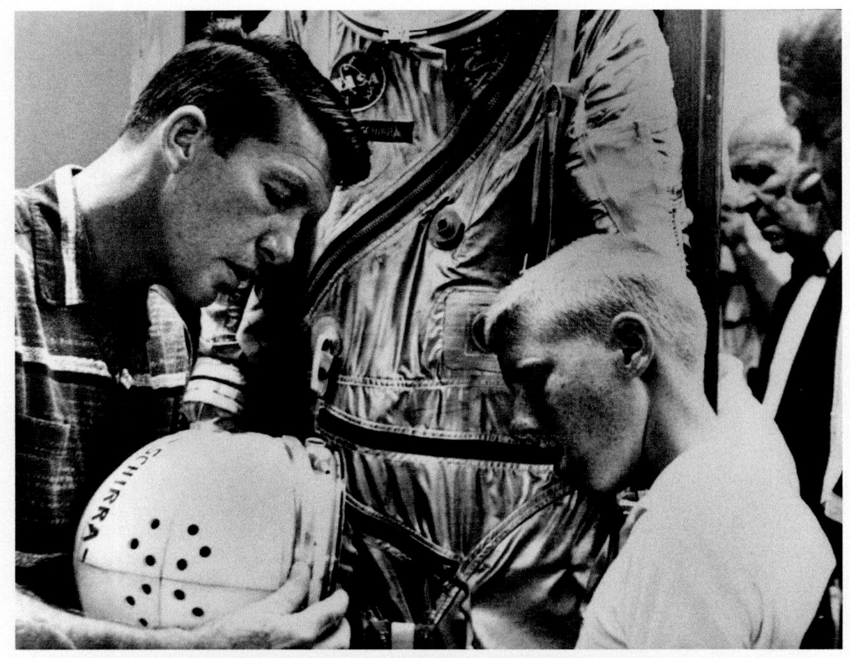

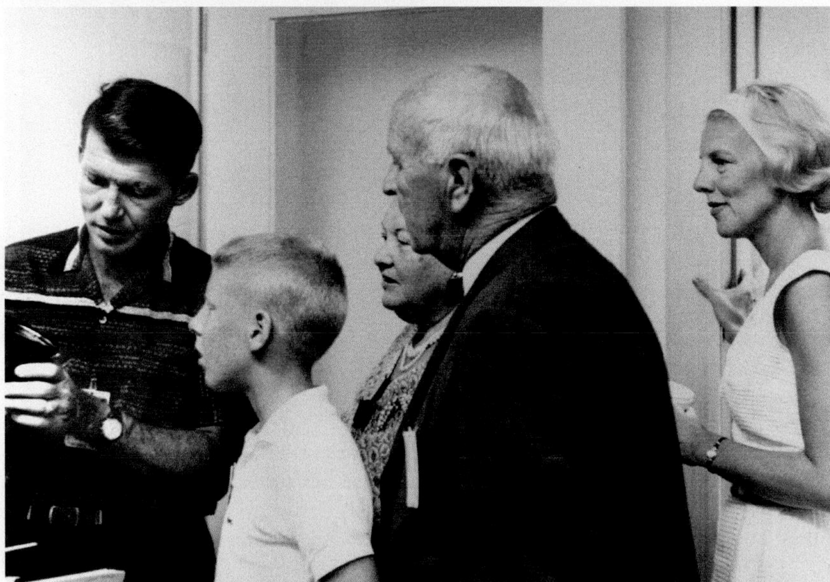

Schirra took time out from his training to show his family the equipment he would use on his flight. (Photos: NASA)

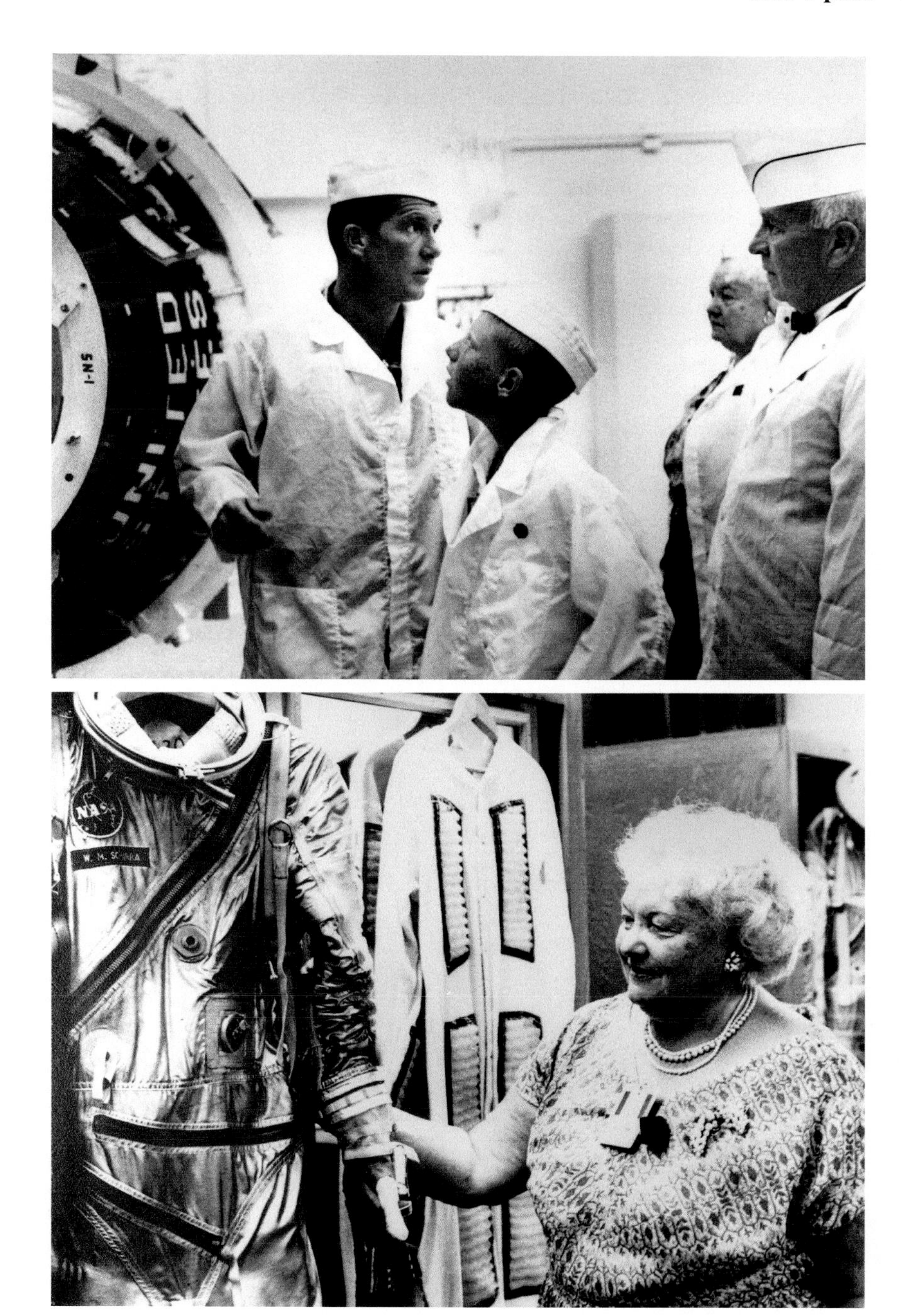

Top: Wearing protective clothing, Schirra allows his family a close-up look at the *Sigma 7* spacecraft. Bottom: Florence Schirra proudly casts an eye over her son's space suit. (Photos: NASA)

The planned scientific program for MA-8 included a flare visibility determination test over Woomera in South Australia. This had been scheduled for Scott Carpenter's earlier MA-7 mission but scrubbed owing to heavy cloud cover over the area. There would also be an electric lamp visibility test over Durban, South Africa on the final orbit involving special lights, equal to three million candlepower, being turned on for three minutes.

Special shingles coated with five different kinds of advanced heat-protection material would also be carried on the flight as a research project for possible use in future heat shields. These would be mounted on the cylindrical upper section of the spacecraft, attached through lamination with the external beryllium shingles. Those participating in this study – AVCO Corporation, NASA Langley Research Center and McDonnell Aircraft Corporation – had been assigned two each of the available nine panels for their respective materials to be tested and analyzed post-flight. Each of the remaining panels had a bonded sample supplied by Emerson Electric, Chance Vought Corporation and General Electric.[21]

Schirra said he would also carry two radiation packs of a new design on behalf of the Goddard Space Flight Center in Maryland in order to obtain information on what might be a possible hazard to future astronauts flying in deep space.

Assembly of Atlas 113-D continues in early August 1962. (Photo: NASA)

The Atlas undergoes final preparation prior to leaving the General Dynamics plant in San Diego. (Photo: NASA)

Wally Schirra autographs the Atlas rocket before it begins its journey to the Cape. (Photo: NASA)

The Atlas rocket departs San Diego. Next stop: Cape Canaveral. (Photo: NASA)

Among other things, Wally Schirra was a camera enthusiast. During his flight, he wanted to take some high quality photographs of recognizable features of the planet as they passed beneath him. He had seen 35 mm photos taken from orbit by John Glenn and Scott Carpenter, but he and Deke Slayton felt that he could improve a lot on those images.

His camera of choice was the Swedish-manufactured Hasselblad 500C, which was held in high regard by professional photographers owing to its superior engineering, craftsmanship and picture quality. Just as John Glenn had done, Schirra purchased the camera himself, this time from a Houston photographic supplier. He gave the camera and lens to NASA, and then the USAF/PanAm camera laboratory worked with him to modify the Hasselblad to enable him to use it within the confines of the Mercury space capsule. This included installing a 100-frame film magazine, painting the original metal casing black to minimize reflections, and mounting an aiming device on its side so that the camera could be used by an astronaut wearing a space helmet and gloves. They also added a planar f/2.8 Zeiss lens with a focal length of 80 mm.

"I sought advice from professional photographers such as Ralph Morse and Carl Mydans of *Life* and Dean Conger and Luis Marden of *National Geographic*," Schirra explained. "I decided that a Hasselblad, with its larger film frame, was more suitable than a 35 mm camera. I had the Hasselblad adapted. A 100 exposure film container was installed, and an easy aiming device was mounted on the side of the camera. Focusing would not be required from the infinity of space, I figured. Finally, I learned how to repair the Hasselblad."[22]

Wally Schirra with the Hasselblad camera he would use on his space flight as Deke Slayton looks on. Left of photo is Paul Backer of McDonnell Aircraft, who trained astronauts in the use of cameras, and on the right is camera specialist Roland ("Red") Williams, an RCA Service Company technician at the USAF Missile Test Center, who modified and greatly lightened the camera for use in space. (Photo: NASA)

A TALE OF TWO NURSES

Prior to the MA-8 flight, and in a process similar to that carried out on the preceding four manned Mercury flights, 23 kits of washed and sterilized instruments – in total 29,900 pounds – were packed into metal boxes, ready to be used in the event of an emergency situation following splashdown. Each of these "Portable Hospital Kits" was made up of a number of cartons strapped together. Several were destined for placement aboard recovery ships stationed under the spacecraft's intended flight path, while others would be shipped to strategic downrange locations around the world. NASA was doing its best to prepare for any contingency in these pioneering flights.

First Lt. Shirley Sineath from Sanford, North Carolina, was the nurse responsible for the packing of these emergency medical kits. "My job for the Mercury flights was Chief Medical Support, Space Flight Operations Sections, Office of the Deputy for Bioastronautics at Headquarters Air Force Missile Test Center, Patrick Air Force Base, Florida," she reflected. "I was there from March 27, 1961 to June 13, 1963."[23] This meant she was involved in every manned flight in the Mercury program.

Together with her roommate Lt. Dee O'Hara from Nampa, Idaho, Sineath had completed the Air Force's flight nurse's course, following which they were both assigned to work with the astronauts in the fledgling Project Mercury. They were under the direction and supervision of Col. George Knauf, M.D., and one of their tasks involved the development of the portable emergency hospital kits for the astronauts.

As recalled by Dee O'Hara, "Colonel Knauf had been tasked by NASA to put together medical support teams from all of the military services – the Army, the Navy, and the Air Force – and these support teams were to consist of surgeons and nurses and people of all disciplines. They would go aboard recovery ships and set up hospitals, should there be a problem upon landing. They were stationed all over the area in case there was a landing that was off-center or wasn't quite where it was supposed to be. What we had to do was to put together these medical kits and everything that people on board the ships would need to treat an injured astronaut."[24]

Initially, the role designated to Lt. O'Hara involved setting up crew quarters and an aeromedical facility in Hangar S at the Cape, and to become the personal nurse to the astronauts, as she explained. "Colonel Knauf wanted someone – a nurse – who would get to know the astronauts so well that she could determine if they were suffering any ailments. Pilots and astronauts are not about to tell a flight surgeon when they're sick, and that's understandable. They are afraid of being grounded, and the flight surgeon is the only one who has that authority over them. So they're not usually very friendly with their flight surgeons."[25]

Dee O'Hara made a deal with the astronauts about any medical problems they might have by promising she would never betray them… "Unless, in my opinion, what they told me would jeopardize them or the mission. In that case, I would have to report it to Dr. Douglas."[26]

For her part, Shirley Sineath was initially involved in the suborbital flights of Alan Shepard and Gus Grissom, as she explained. "My main task was preparing nine one-man Portable Hospital Kits, each weighing 1,933 pounds. Seven of these were placed aboard the recovery ships under the astronauts' intended flight path, one at a forward medical facility at Canaveral, and one at Grand Bahama Island where the astronauts were to be taken post-flight. These kits also included bundles of sterilized instruments for use in the portable hospitals."[27] As she explained to reporters back in 1961, "The Portable Hospital Kit contains everything from bandages to anesthesia machines and resuscitators. They carry everything necessary for an emergency operation on a ship or on one of the land medical areas. Each package has been checked and double-checked to make certain nothing is left out. One missing item could mean the difference between life and death for the astronaut."[28]

The medical kits were only one of many important pre-flight duties Shirley Sineath carried out. During each manned launch, Sineath served as the surgical nurse and Dee O'Hara was the intensive care nurse at the Cape's forward eight-room medical facility. This was within a reinforced concrete structure, designed to protect its occupants in the event of an explosion on the nearby launch pad. Two surgeons and an anesthesiologist were among 55 medical personnel also on duty, ready to provide emergency care to the astronaut if a catastrophic malfunction occurred during the launch phase. By procedure, anyone else injured in the mishap would be cared for by registered nurses from the Pan American World Airways medical-nursing team.

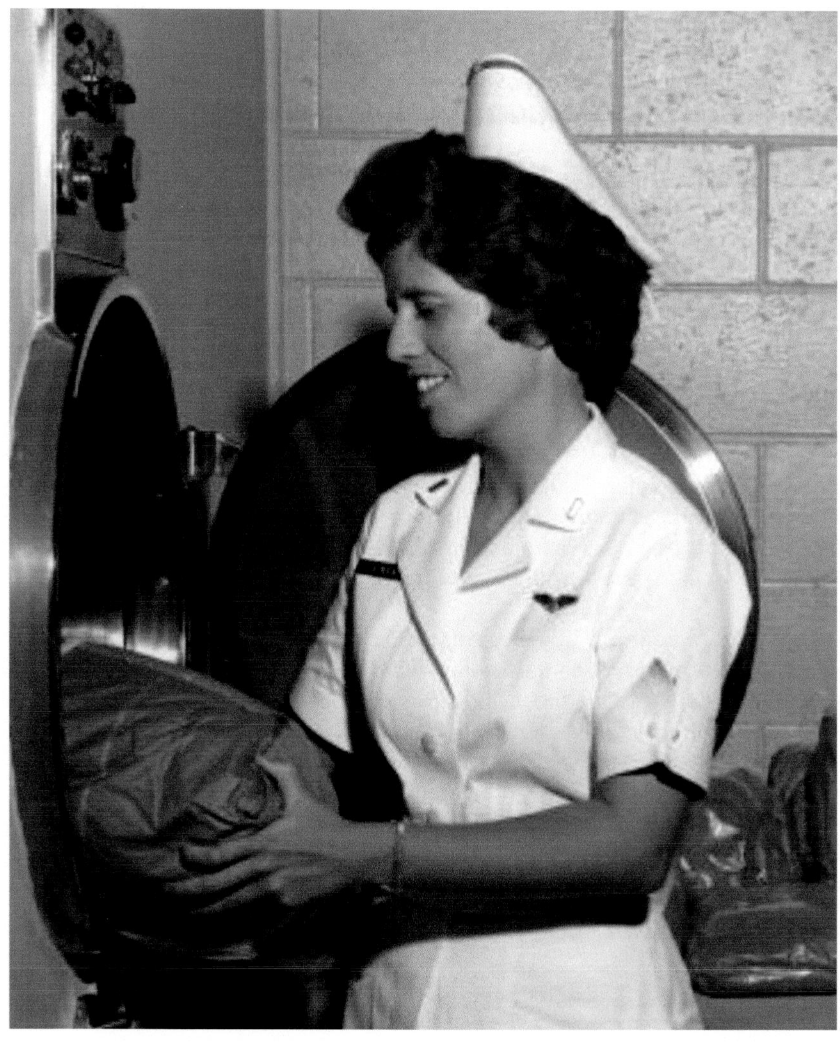

Nurse Lt. Shirley Sineath prepares a medical package to be used in connection with the recovery of Wally Schirra after his MA-8 mission. (Photo: NASA)

In 2003, Dee O'Hara was asked for her thoughts concerning Wally Schirra. With a smile and a laugh, she responded, "Now, what can one say about Wally Schirra – that Wally hasn't said before! There was no question; he was the personality of the group. With every quip or pun, it didn't matter what you said to Wally; you simply could not get ahead of him – ever. He's extremely quick on his feet; very, very sharp, very witty, and deadly serious when it comes to work. Boy, is he serious! And you don't mess with him on his flight. He was the commander of that flight and you better know that he was in charge. You had to march to a certain beat. He was very strict in his work ethics and his crew had to be the same. He did not tolerate any mistakes. But Wally, he's wonderful. He could take any sort of a situation and make it so much fun. You could always hear him laughing coming down the hall and all over the place. He's just a delightful guy."[29]

Wally himself would have agreed with those sentiments. Indeed, as he stated with conviction in his memoirs, "Humor was an essential ingredient of our existence at Cape Canaveral. It provided relief from very long hours of very serious preparation for flights into the unknown. More important, it palliated the frustrations of those early days in Mercury."[30]

On 29 June 1963, two weeks after Shirley Sineath had completed her work on Project Mercury, she married John ("Whipper") Watson from the Canadian Air Force, who was also involved in the Mercury flights as navigator in a CF-100 Canadian jet fighter, flying downrange to track launch vehicles after lift-off. "My discharge date from the Air Force was January 24, 1964," she recalled. "After discharge I came to Canada as a housewife and continued my career in nursing until retirement in 2004. My three goals in life were to be a nurse, to be in the Air Force, and to be a Flight Nurse. Never did I think I would be selected to be in the space program. I was lucky enough to be in the right places at the right time."

Asked for her opinion of Wally Schirra, Shirley felt much the same as Dee O'Hara. "Although I was not in direct contact with Wally, my impression of him was of a very professional and family-oriented man. He was very dedicated to his work as a Mercury astronaut and took the position extremely seriously."[31]

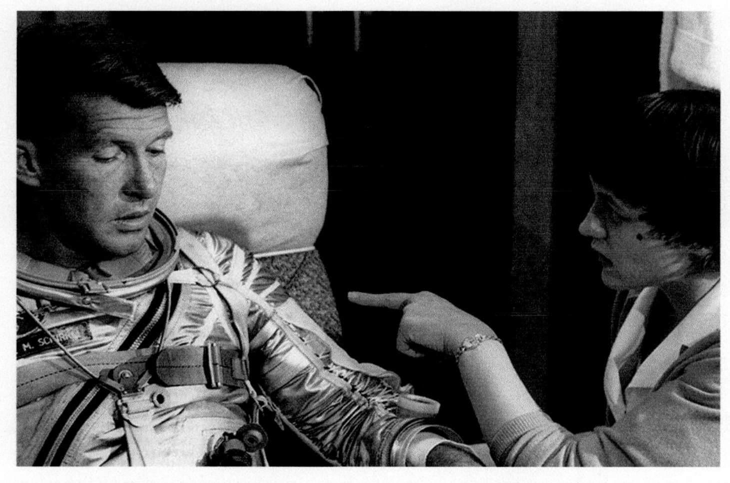

Captured in a serious moment; Dee O'Hara talks with Wally Schirra. (Photo: NASA)

Dee O'Hara joins in the gag by hugging the "urine specimen" jar. (Photo: NASA)

Professional and serious they might have been when it came to their work, but the astronauts' humor extended to playing jokes or "gotchas" on those they worked with – and Dee O'Hara was no exception. "Oh yes. For instance, they used to hide their urine samples from me. I'd find them in the shower stalls, the air conditioner, and the heater – anywhere! One time that great practical joker Wally Schirra was asked to provide a urine sample and he brought it to me in this gigantic container the size of a water cooler!"[32]

Wally Schirra recalled the gag well, and said he and Gordon Cooper had worked it out between them. "Early one morning we filled a five-gallon bottle with warm water, figuring it would cool to body temperature by the time she arrived. We added a bit of iodine to give it color, laundry soap to make it foamy, and put the bottle on Dee's desk. I tagged the bottle, writing the time of delivery in Greenwich Mean Time, and I attached a bunch of lollipops. Gordo and I were like little kids, peeking with glee around the corner of the doorway, when we spotted Dee first stop in horror, and then burst out laughing."

On a more serious note, Schirra said that Dee was "like a sister to us, and a mother to our children. Dee could take care of us better than any of the doctors could. We allowed no one but Dee to draw our blood, not even the doctors, for fear that they would collapse a vein."[33]

This always amused Dee. "The guys got on this thing about nobody doing it but me. What these guys didn't realize was that I was really out of practice, and I was probably the worst person in the world to do it. God was good to me – I never missed a vein!"[34]

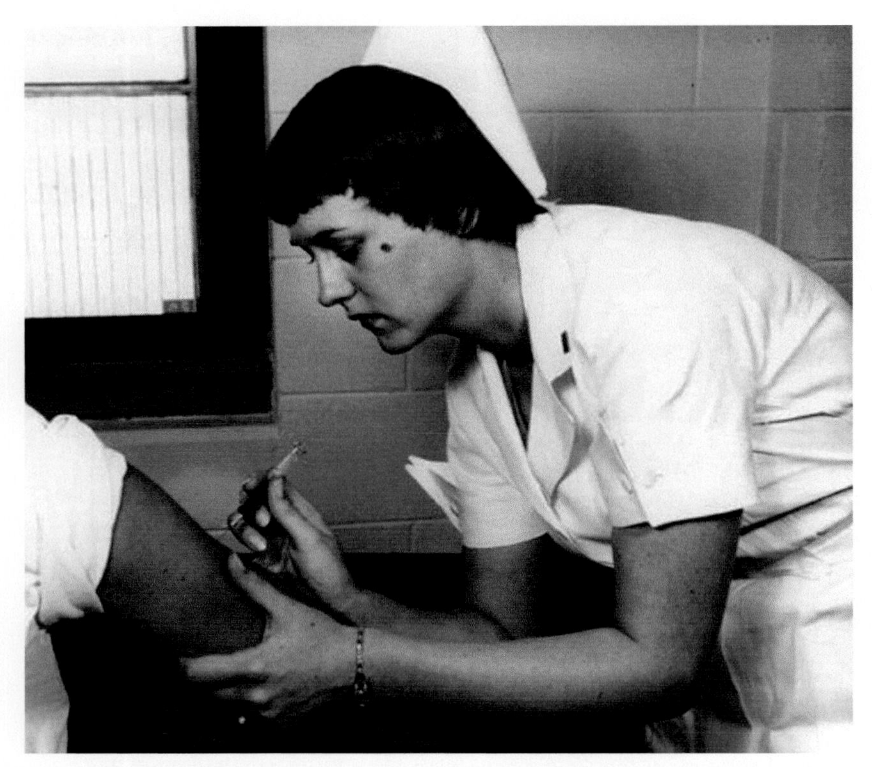

Dee showing that she can actually do it right, despite her pessimism. (Photo: NASA)

Schirra continues to train hard for his upcoming mission. (Photo: NASA)

GETTING READY TO FLY

Another big fan of Wally Schirra was NASA's ebullient head of Public Affairs, Paul Haney, who gave the author another insight into the astronaut's wry sense of humor:

> Of all the early astronauts, I found Wally Schirra most interesting and spontaneously funny. He was also spontaneously serious at work. But he could take or deal a joke with the best, which always lightened otherwise heavy loads. We used to trade double and triple entendres to keep from falling asleep in some briefings. An Annapolis graduate, Schirra married into a Navy family which produced several Chiefs of Naval Operations. The Navy has a passion for naming commands or theaters by initials. A Chief of Naval Operations, for example, is a CNO. The commander of the Atlantic Fleet is CinCLant.
>
> Schirra and I were going down a receiving line at a social function in Honolulu. One of the honored guests was the Red Dean of Canterbury, there in brilliant red vestments, representing Presbyterians everywhere. I heard Wally greet the churchman with something appropriate like, Your Eminence. And the cleric responded by recognizing Wally as one of the American astronauts. Then, chatting the man up a little, Schirra said, "You're the head of the Presbyterian Church throughout the Pacific?"
>
> The Dean smiled and nodded. Then Wally hit him with this, "That makes you ComPresPac, right?"[35]

NASA's Paul Haney, a member of Wally's "Turtle Club." (Photo: NASA)

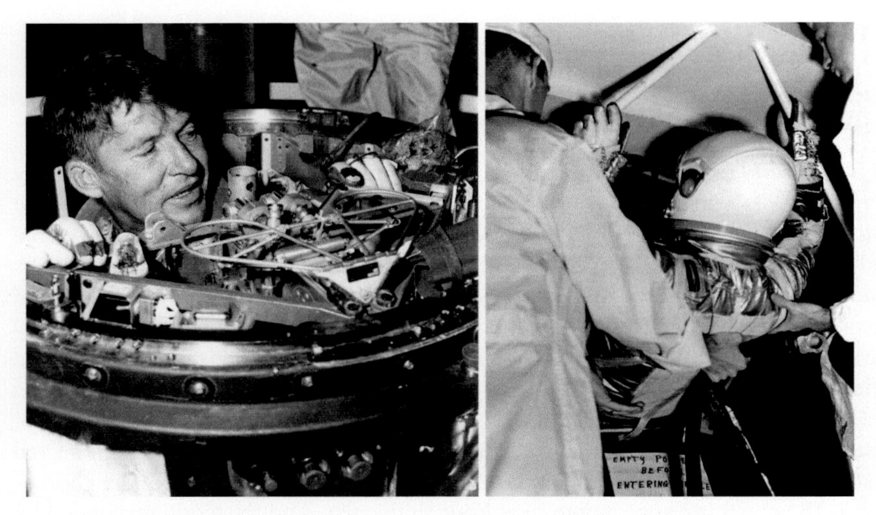

Practicing egress and escape techniques. (Photos: NASA)

As they had done prior to the previous Mercury missions, the Soviet Union once again sent shockwaves across the United States with the announcement of yet another space "first" as NASA was preparing to send Wally Schirra on his six orbit flight. On Saturday, 11 August, Radio Moscow announced that at 11:30 a.m. that day (Moscow time), the Vostok-3 spacecraft had been launched into Earth orbit carrying cosmonaut Maj. Andrian Nikolayev, who was completing one revolution of the planet every 88.5 minutes.

While Western observers pondered how long Nikolayev would remain in orbit, there was an even greater surprise 24 hours later when Radio Moscow interrupted its normal programming to issue a further announcement. The startling news informed listeners that at 11:02 a.m. Moscow time, a second cosmonaut had been shot into orbit from the same launch pad as the one used the previous day. As Vostok-4 slipped into orbit with cosmonaut Pavel Popovich aboard, history was being made with two pilots in space at the same time. There were rumors circulating that the two spacecraft might link up in orbit, but this was never the case because their orbits were dissimilar and they were not equipped for such a task. However, Popovich did reveal at a post-flight foreign press conference that he knew where Vostok-3 would be in relation to his own craft once he was in orbit, and did manage to spot the other vehicle which looked "something like a very small moon in the distance."

Nikolayev would eventually complete 64 orbits and Popovich 48 before both men returned safely on 15 August to conclude their historic dual mission.

At a press conference held on 16 August, Dr. Brainerd Holmes, NASA's director of space flight programs, suggested that the MA-8 mission would likely take place in the middle or end of September. In view of the recent Soviet space spectacular, Holmes responded to questions by stating that even if everything went perfectly on the flight,

Soviet cosmonauts Andrian Nikolayev and Pavel Popovich. (Photo: Author's collection)

Schirra would not be permitted to fly more than the scheduled six orbits. Speaking on behalf of the space agency, he made it clear that even though Popovich and Nikolayev had spent three and four days in space respectively, there would be no extension to the forthcoming Mercury flight because the spacecraft was not large enough to contain the oxygen and other supplies for extended orbital flights. "The mission is scheduled for recovery near Midway Island after six orbits," he stated, "and that is the way it will be flown."

Holmes added that the recovery force of ships and aircraft for MA-8 would have to be "half again as big" as that for the two previous manned orbital flights. This was due to the necessity of maintaining recovery watches from the mid-Atlantic and Bermuda waters through the Caribbean and out across the Pacific, in case the spacecraft should come down short of the assigned six orbits.

Pressed further about the Soviet tandem flight, Holmes said the United States was unable and unwilling at that time to orbit two spacecraft simultaneously, as NASA only had the one Mercury-Atlas launch pad. Similarly, the Soviets only had the one Vostok launch pad, but the inordinate haste involved in setting up and launching two manned space missions on consecutive days was a risk NASA would not even contemplate. "I don't think we would choose to do it with the Atlas booster even if we had twin launching capability."[36]

Dr. Brainerd Holmes (left) in conversation with Dr. Wernher von Braun. In center is Nicholas Golovin, a physicist and member of the president's Science Advisory Committee. (Photo: NASA)

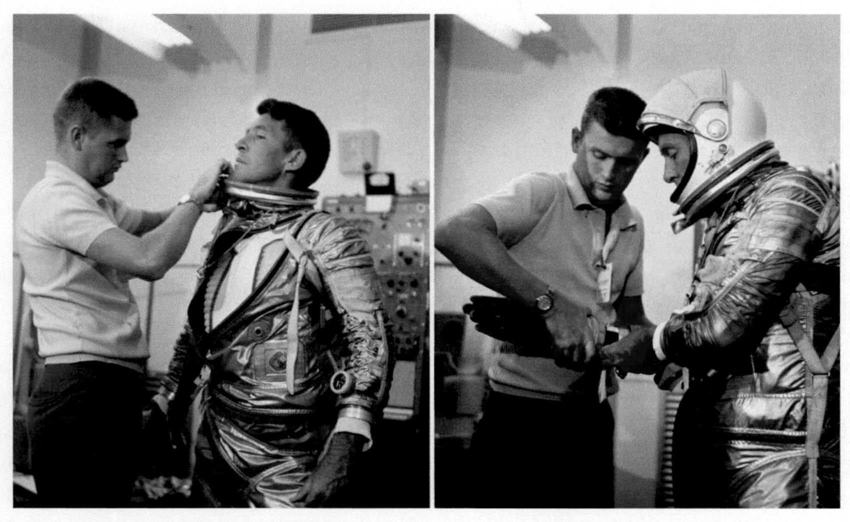

Suit technician Al Rochford and Wally Schirra practice the routine of donning his space suit and helmet. (Photos: NASA)

Prior to the installation of *Sigma 7* atop Atlas 113-D, the rocket underwent a planned static test firing on Pad 14 to evaluate recent modifications. (Photo: NASA)

On Monday, 10 September, the bell-shaped spacecraft No. 16, *Sigma 7*, was installed on top of the towering Atlas LV-3B rocket at Cape Canaveral. In a delicate operation, the 1½-ton spacecraft had been hoisted to the top of the gantry at Launch Complex 14, where technicians were waiting to carry out the critical task of mating it to the booster.

The previous day, Atlas 113-D had successfully undergone the static firing required by the Air Force to test a new fuel feeding system. With the vehicle firmly attached on the pad, the main engines had successfully fired for 11 seconds.

At this time, NASA announced a further three-day postponement in the launch date, to Friday, 28 September. However, the Atlas rocket suffered a leak through a seam in its fuel tank which had to be repaired, and any hopes of a September launch were lost. NASA issued a statement that the launch would take place no earlier than 3 October.

Meanwhile, the recovery forces had taken up their positions. Five ships under the command of Rear Admiral Charles A. Buchanan, Commander Task Force 130, would make up the prime recovery force north-east of Midway Island. Given the possibility that

Sigma 7 being raised to the top of the gantry for mating to the Atlas rocket. (Photo: NASA)

Schirra watches intently as the spacecraft is lowered into position atop Atlas 113-D. (Photos: NASA)

the flight might only last up to three orbits, more than 20 ships had been deployed to the Atlantic Ocean recovery areas, under the command of Rear Admiral Howard G. Bowen, Destroyer Flotilla Four. As well, some 100 aircraft around the world were on standby, ready to be called into action in the event of the flight being aborted early, or an emergency landing.

The Manned Space Flight Network which would monitor the MA-8 mission also comprised five tracking ships spread across the vastness of the Pacific Ocean: USNS *Rose Knot* (T-AGM 14) stationed off Guam; USNS *Huntsville* (T-AGM 7), USNS *Watertown* (T-AGM 6), and USNS *American Mariner* (T-AGM 12) all of which were in the vicinity of Midway Island; USNS *Range Tracker* (T-AGM 1) off Hawaii. And in addition USNS *Coastal Sentry* (T-AGM 15) would serve as the Indian Ocean tracking and communications ship.

In keeping with normal practice, the other six Mercury astronauts would participate in duties connected with the MA-8 mission. At Cape Canaveral, Deke Slayton would serve as Capsule Communicator (CapCom), while backup pilot Gordon Cooper would also involve himself at the Cape. Alan Shepard was aboard the Pacific Command Ship USNS *Rose Knot* while Gus Grissom was stationed on Kauai Island, Hawaii. John Glenn was at the tracking station at Point Arguello, California, and Scott Carpenter was at the tracking station at Guaymas in Mexico.[37]

All was in readiness for NASA's next big step in exploring, defining and conquering the new frontier of space.

Gordon Cooper in discussion with Schirra during pad training aboard *Sigma 7*. (Photo: NASA)

Deke Slayton in the Mercury Control Center to serve as Cape Canaveral CapCom during the MA-8 mission discusses the countdown with Gordon Cooper. (Photo: NASA)

REFERENCES

1. M. Scott Carpenter, L. Gordon Cooper, Jr., John H. Glenn, Jr., Virgil I. Grissom, Walter M. Schirra, Jr., Alan B. Shepard, Jr., and Donald K. Slayton, *We Seven*, Simon and Schuster, New York, NY, 1962
2. Walter M. Schirra, Jr. with Richard N. Billings, *Schirra's Space*, Quinlan Press, Boston, MA, 1988, pg. 60
3. Transcript of introduction of NASA's Mercury astronauts, Dolley Madison House, New York, NY, 9 April 1959
4. Walter M. Schirra, Jr. with Richard N. Billings, *Schirra's Space*, Quinlan Press, Boston, MA, 1988
5. Robert B. Voas, NASA Space Task Group, "Project Mercury Astronaut Training," from *Report of a Working Group Conference, Armed Forces – NRC Committee on Bio-Astronautics*, Publication 873, Washington, DC, 1961
6. M. Scott Carpenter, L. Gordon Cooper, Jr., John H. Glenn, Jr., Virgil I. Grissom, Walter M. Schirra, Jr., Alan B. Shepard, Jr., and Donald K. Slayton, *We Seven*, Simon and Schuster, New York, NY, 1962, pg. 118
7. Shirley Thomas, chapter "Alan B. Shepard," from *Men of Space, Vol. 3*, Chilton Company, New York, NY, 1961, pp. 195–196
8. M. Scott Carpenter, L. Gordon Cooper, Jr., John H. Glenn, Jr., Virgil I. Grissom, Walter M. Schirra, Jr., Alan B. Shepard, Jr., and Donald K. Slayton, *We Seven*, Simon and Schuster, New York, NY, 1962, pg. 92
9. Walter M. Schirra, Jr. with Richard N. Billings, *Schirra's Space*, Quinlan Press, Boston, MA, 1988
10. *Ibid*
11. John H. Boynton interviewed by Rebecca Wright for NASA JSC Oral History program, Houston, TX, 6 March 2009
12. *Newsweek* magazine unaccredited article, "Schirra in Orbit," issue 24 September 1962
13. Walter M. Schirra, Jr. with Richard N. Billings, *Schirra's Space*, Quinlan Press, Boston, MA, 1988
14. NASA *Space News Roundup*, "Schirra To Pilot Six-Orbit Mission in September," Manned Spacecraft Center, Houston, TX, 11 July 1962, pg. 1
15. Wes Oleszewski, "Sigma 7 – Fifty Years Ago Today," *Aero News Network*, 3 October 2012, online at: *http://www.aero-news.net/index.cfm?do=main.textpost&id=59a57630-958a-480b-9f0a-8658835f4abd*
16. Manfred (Dutch) von Ehrenfried, *The Birth of NASA*, Springer-Praxis Publications, New York, NY, 2016
17. Jerry Bledsoe, *Esquire* magazine article, "Down From Glory," issue January 1972, pg. 83
18. NASA *Space News Roundup*, "Schirra Announces Spacecraft Changes Via Telstar," Manned Spacecraft Center, Houston, TX, 25 July 1962, pg. 1
19. NASA *Space News Roundup*, "MA-8 Readies For Six Earth Orbits," Manned Spacecraft Center, Houston, TX, 3 October 1962, pp. 1–2
20. David Baker, Ph.D., *The History of Manned Space Flight*, Crown Publishers, New York, NY, 1981

21. NASA *Space News Roundup*, "Four Experiments Will Be Carried Out During Flight," Manned Spacecraft Center, Houston, TX, 3 October 1962, pg. 3
22. Walter M. Schirra, Jr. with Richard N. Billings, *Schirra's Space*, Quinlan Press, Boston, MA, 1988, pg. 80
23. Shirley Sineath Watson email correspondence with Colin Burgess, 23–28 July 2015
24. Dee O'Hara interview with Rebecca Wright for NASA JSC Oral History program, Mountain View, CA, 23 April 2002, amended for author by Dee O'Hara, 23 July 2015
25. *Ibid*
26. Colin Burgess, "Astronaut's Nurse: An Interview With Dee O'Hara," BIS *Spaceflight* magazine, Vol. 43, No. 7, July 2001
27. Shirley Sineath Watson email correspondence with Colin Burgess, 23–28 July 2015
28. *Spartanburg Herald* (NC) newspaper, unaccredited article, "'Space Hospital' Readied by Nurse," issue 1 May 1961, pg. 3
29. Dee O'Hara interview with Colin Burgess and Francis French, San Diego, CA, 18 January 2003
30. Wally Schirra with Richard N. Billings, *Schirra's Space*, Naval Institute Press, Annapolis, MD, 1988
31. Shirley Sineath Watson email correspondence with Colin Burgess, 23–28 July 2015
32. Dee O'Hara interview with Colin Burgess and Francis French, San Diego, CA, 18 January 2003
33. Wally Schirra with Richard N. Billings, *Schirra's Space*, Naval Institute Press, Annapolis, MD, 1988
34. Francis French and Colin Burgess, *Into That Silent Sea: Trailblazers of the Space Era 1961–1965*, University of Nebraska Press, Lincoln, NE, 2007
35. Paul Haney email correspondence with Colin Burgess, 21 September 2003
36. The *Spokesman-Review* newspaper (Spokane, Washington), "Delay in Mercury Launch Indicated," issue 17 August 1962, pg. 2
37. NASA *Space News Roundup*, "MA-8 Readies For Six Earth Orbits," Manned Spacecraft Center, Houston, TX, 3 October 1962, pg. 2

3

The art of space flight

Over the decades a large number of people touched the life of Wally Schirra in so many ways, and he had a surprising recollection for people he had met, quite often attaching a pun to their names. One person who sometimes shared in Wally's pressure-relieving "gotchas" during the early 1960s was a wonderfully talented woman who will be forever (and rightfully) linked to the Mercury space program. Her name was Cecelia Rose Bibby, better known to everyone as Cece, and she was a much loved and high-spirited graphic artist with a feisty sense of humor.

In 1962, while working for the Chrysler Corporation, Cece was invited by John Glenn to hand-paint the *Friendship 7* logo on the exterior of his Mercury spacecraft. Her colorful artwork proved such a hit with the astronauts that she went on to create similarly striking designs at their request for Scott Carpenter's *Aurora 7* and Wally Schirra's *Sigma 7* capsules. Due to a relocation move back to California, Schirra's capsule would be the last Mercury spacecraft to bear Cece's memorable artwork.

It was by no means an easy task. In order to paint the insignia on each of the corrugated capsules, Cece became the first (and only) woman to ascend the Mercury launch gantry and work inside a busy shielded platform that temporarily surrounded a spacecraft undergoing preparation for flight. This caused her to come in contact with an abrasive capsule chief/pad leader named Guenter Wendt, who made his feelings abundantly clear about having any females intruding into his work zone.

REALIZING A CHILHOOD AMBITION

At the age of nine, along with her brother Frank (known as "Bud") Cece Bibby was thrust into the Masonic Home for Children in Covina, California by their widowed mother Mary, who, after the loss of her husband Frank Bibby six years earlier, had struggled with alcoholism. The two children remained as wards at the Home until of legal age, and never saw their mother again.

C. Burgess, *Sigma 7*, Springer Praxis Books, DOI 10.1007/978-3-319-27983-1_3

As Cece told interviewer Lawrence McGlynn in 2005, becoming an artist wasn't something that she consciously planned. "As a child that's all I ever wanted to be. I used my crayons on everything, probably even on walls. I cannot remember a time when I wanted to be anything else. The Home saw to it that I had the art supplies I needed. Not all the ones I wanted, but what I needed."

After graduating from high school in 1946, Cece left the Masonic Home to enter art school – "I hated to leave. It was my home in every sense of the word." She began to support herself by taking a job as an operator with the local telephone company, and during the Korean War served as an ensign in the U.S. Navy. A move to the East Coast came when Cece was hired as a draftsman for RCA's Missile Division at Patrick Air Force Base, "doing 'as built' drawings for the downrange missile tracking sites." She was then transferred to RCA's publications department, "where we did all kinds of artwork such as engineers' ideas of space."

Her next move was to the Aerospace Corporation. "Most of the people there had Ph.D.'s except for me," she reflected for the interview. "I did illustrations based on their ideas about the future of space. Ideas like [space] hotels, space shuttles or space stations. These ideas were figments of someone's imagination, and I would convert them to pictures. What they would do is sit down and talk to me and tell me about their idea and I would then draw it in color."

Cece Bibby posing in front of a Lockheed U-2 spyplane at Patrick Air Force Base. (Photo: Cece Bibby/Lawrence McGlynn)

In 1959, Chrysler hired Cece as a contract artist at the NASA publications office, Cape Canaveral, where she worked in the NASA Administration Building across the street from the astronauts' office.

Back then, Chrysler was one of many bidders for the technical support contract which was to supply engineering and public relations writers, editors and artists to NASA. "Chrysler won the contract and hired people to staff the department," Cece recalled. "We were not going to work at the Chrysler building but were physically assigned to the Administration Building at the Cape. Although Chrysler would be signing our pay checks, we would be under the direct supervision of a NASA civil servant. Any NASA department could request artwork, technical publications, etc., from our group."

Cece admits she was just as enthralled with space as anyone else who worked at the Cape. "Most of us there stopped everything we were doing to watch a launch. If a launch took place at night we would congregate on the beach to watch. Night launches were spectacular. I lived with three other young women in a house right up on the beach. Next door to us were five Air Force officers. When we knew there was going to be a night launch we'd all get together down on the beach. We'd dig a hole in the sand and build a fire in it, put a grate over it, and have a hot dog roast. We'd wait around all night for the launch. Even if it was put on hold, we'd sit there until early morning. On the other side of our house there was a theodolite tracking station, manned by camera operators, and they would let us know the status of the launch. We'd take them over some hot dogs and Cokes."

Once the two suborbital Mercury missions of Alan Shepard and Gus Grissom had been completed, the next phase of the program was an orbital flight by John Glenn. The popular Marine was keen to follow the tradition set by Shepard of naming his spacecraft, but he felt the simple stenciled block names on the side of the two previous spacecraft didn't really express the spirit of the program. That was how Cece became involved.

> When it came time for John's flight, he told someone in the astronauts' office that he didn't want the name put on by stencil and he didn't want it in block letters ... he wanted it in script. The astronauts' office called over to our arts department and talked with our boss, who made a fast trip over to talk with John Glenn. The boss came back to the arts department and said that John wanted the name done in script. He also said that a man would have poor handwriting so he wanted me to do the job because – as a woman – I might have better handwriting. Big compliment. So, I made about three designs. The boss took them over to John Glenn's office. John selected the one he wanted and said he wanted the artist to apply the design to the capsule. My boss said that a stencil could be cut and one of the guys could apply the name that way. Wrong thing to say. I found out about the conversation later from John and it went something like this:
>
> John: I want the artist who designed that to put it on by hand.
> Boss: Well, that's a woman.
> John: So?
> Boss: She'd have to go out to the launch pad and up to the top of the gantry.
> John: Is she handicapped in some way?
> Boss: Well, she's a woman (to my boss being a woman was a handicap).

John: Is she afraid of heights?
Boss: I don't know ... but she's a woman.
John: Why don't you find out from her whether she has some objection to going up to the top of the gantry to paint this for me? Let me know what she says.

So, the boss came back to the arts department, threw the designs on my drawing board and said that John Glenn was requesting that I personally go out to the pad to hand paint the design he had chosen. I couldn't figure out why his face was so red ... because I had no idea that the boss hadn't got his way. John told me all about the conversation later.

Cece painting the *Friendship 7* logo on John Glenn's spacecraft. (Photo: NASA)

Cece would find painting on the corrugated surface was interesting, but not all that difficult. "I had to be really careful though, about how I transferred the designs to the surface of the capsule. Normally I would have used a graphite type transfer paper but that has lead in it and there was a fear that the lead could cause cracks. I chalked the back of my design and very gently traced it onto the surface. I can't remember how long it took me to paint the designs. I had to base coat and allow it to dry before applying the final coat and I don't remember how many coats I put on."

Following the flight of *Friendship 7*, Cece was able to see the damage caused to her artwork by the ferocious heat of re-entry. Nevertheless, to her delight, the logo was still readily visible on the hull of the historic spacecraft. "I wasn't pleased about the paint burning off after all of my work, but we had figured it probably would. The heat upon re-entry was tremendous so we'd have been surprised if any of the designs had managed to make it through unscathed."

John Glenn being squeezed into *Friendship 7* on launch day. Cece's logo is clearly evident. (Photo: NASA)

The fact that Cece had painted the *Friendship 7* logo on Glenn's spacecraft certainly attracted its fair share of media attention, making her something of a minor Mercury celebrity, often with humorous results. "I had a couple of people contact me a few years ago wanting to know whether I was the one who had painted the crack on Gus Grissom's *Liberty Bell 7*. I assured them I didn't do that. One of these guys said he was at a book signing for John Glenn, and he asked John about who painted the crack on *Liberty Bell*. He said that John got a grin on his face and answered, 'I think maybe Cece Bibby did that.' John had forgotten that his was the very first job I did for the astronauts and it wasn't possible that I got anywhere near the *Liberty Bell* capsule. I can't seem to convince people that I didn't paint that darned crack on Gus's capsule. Oh well…"

Cece recalls that Glenn also took with him on his flight twelve gold medallions bearing the engraved *Friendship 7* insignia. "John did have the idea that he would like to have some kind of memento to give to his wife and children, to his secretary, and to the astronaut nurse, Dee O'Hara. So, we came up with the idea of doing gold charms [medallions] with *Friendship 7* engraved on them. I had a jeweler friend of mine make up a dozen of these charms and John carried them on his flight. I did the same for Wally Schirra and Scott Carpenter… they each ordered a dozen of the charms and carried them with them on their flights. So, there are really only twelve charms for each of the flights. I think these might have been the precursors of the patches. People knew about the charms and wanted some memento of their own. Some enterprising salesman probably approached someone at NASA and the patch idea was born."

One of Cece's Mercury charm bracelets. (Photo: Cece Bibby)

Gus Grissom: a test pilot with the need for speed. (Photo: NASA)

Cece says she enjoyed working with the Mercury astronauts, both as a group and individually, but she sometimes found that being so closely associated with them and their often egocentric ways was difficult:

> Alan Shepard … I just couldn't warm up to his arrogant attitude. He came swaggering into the arts department one day and all of the other artists were falling all over him … they were all guys and into hero-worship. Al came over to my drawing board and said, 'I want to drive your car.' It was a demand. I told him I didn't think so. Just a few days previously he had a rental car out on the beach and blew up the engine on it. He just threw the keys in the car and left it. And he expected me to jump at the chance to have him drive my car! Like he was doing me a big favor. He said, 'Well, you've let Wally and Gordo drive it.' My response was, 'Yes, I do let my friends drive it.' I thought the implication was quite clear. Al's eyes were always bulgy and they bulged out even more at that comment. He turned on his heel and swaggered out of the arts department. The other artists were all over me about how rude I had been to Shepard and wanted to know how I could have done that. I told them about the blown engine on the rental car and that I wasn't about to have that happen to mine. My car had a BMW 507 engine in it and I didn't intend to replace it.
>
> Gus Grissom: I liked Gus, but he wasn't all that easy to know. He did give me a tour of Houston and the Clear Lake area in his Corvette. I might as well have been in a whirlwind … we went about 80 or 90 mph through downtown Houston. I lived to tell about it but it aged me a good ten years. I should have known what kind of tour I would get of Houston. When Gus brought me back to my apartment complex he said, 'Well, what do you think of Houston?' I told him I'd have to let him know after I had seen it. He just drove off laughing.

PROBLEMS ON THE LAUNCH GANTRY

As a result of Cece painting the artwork on the side of the *Friendship 7* spacecraft, she was invited by Scott Carpenter to do the same for his *Aurora 7*. Having recently been assigned the second orbital flight following the unexpected standing-down of Deke Slayton due to a minor heart ailment, Carpenter was in the midst of a heavy training schedule for the upcoming MA-7 mission. But having re-christened the spacecraft *Aurora 7* instead of Slayton's *Delta 7* he wanted Cece to emulate the wonderful work she had done on Glenn's capsule. She was naturally thrilled to have been handed this new assignment, and there is an interesting story attached to the *Aurora* logo. It came with a chance to do some practical experimentation.

"Scott's flight was the first flight where we actually ran a paint test. I don't know what paint was used on the first two flights – Al and Gus – except that it came out of a spray can, as far as I know. I used paint that NASA gave me for John's flight."

NASA had discovered that the effect on exterior paint was something they hadn't thought about until Glenn's flight, and only after they observed that the ferocious heat of re-entry had burned off most of the paint.

Cece was recruited by Scott Carpenter to paint the *Aurora 7* logo on his spacecraft. (Photo: NASA)

Scott Carpenter inspects Cece's artwork. (Photo: NASA)

As Cece explained:

> That made them realize that they needed to do some paint tests. Scott's flight had three different kinds of paint; one of which came from Pittsburgh Paint, I think another might have come from Glidden, but I'm not sure about that. One could have been from Sherwin/Williams.
>
> I was an avid sports car enthusiast, having had a couple of MG-As, a TR-3, and one of the first Corvettes. I used to go to sports car races throughout the south, when time permitted. I'd go to the races at Sebring, Florida, and work on the timing crew. A friend of mine, Joe Lane, was the head timer for the Sebring races and he also worked for Pittsburgh Paint. Anyway, Joe asked about the paint I had used on John Glenn's flight and the paint test just sort of grew from there. He talked to me about the company making a special paint that might be impervious to re-entry heat and I suggested a test on Scotty's flight. I obtained paint from Joe's company and someone got paint from the two other companies. I divided up the letters in Aurora; I used one kind of paint on 'AU', another on 'RO', and the third on 'RA'.
>
> If you look at the color shot of the whole design you'll see the reason for doing the number 7 in blue. With the name in white and the auroras in reds, yellows and even orange, it just made sense to do the number in blue. After Scotty's flight we were really glad I had made the number in blue and that was for ideological reasons.
>
> The Russians/Soviet Union made a big thing out of the name *Aurora*, giving a lot of play in their papers (and internationally) about the fact that the first ship to fire a shot in their revolution against the Tsar was named *Aurora*. We were just glad I hadn't painted the number 7 in red for that particular name. We hadn't even given the USSR's ship *Aurora* a thought, so the choice of color, before Scotty's flight, hadn't even come into play. As I said, it was just the combination of colors that mattered, and I was sure glad of it when the Russians began their crowing about the name. Let's face it; the USSR had a thing about the color red.

But choosing a color scheme was not the biggest problem that Cece faced on the launch gantry. There was her ongoing conflict with pad leader Guenter Wendt. The first time she made an appearance at the top of the wind-affected gantry to paint John's Glenn's *Friendship 7* insignia under difficult-enough conditions she was confronted by Wendt, who curtly informed her that women weren't allowed up there and that she was to leave immediately. As she told interviewer and friend Lawrence McGlynn:

> I was there legitimately, doing my job, and was out of the way on the side of the capsule. No, Guenter's complaint was that a woman was up in the room on top of the gantry, God forbid, in what Guenter considered his territory. We couldn't have females up in an all-male area. I told him he'd have to take it up with John Glenn and I went ahead and did my job. He was always trying to boss me around when I was up there trying to paint on a capsule. He just wouldn't let up, picking at me for any reason he could find, and he wasn't even my boss. My boss was bad enough; I didn't need Guenter playing proxy boss. And saying I was always in the way.

It was very difficult to work on the project when you were being subjected to ridicule. That, coupled with the fact that the capsule was being checked out and subject to movement or tests caused the whole project to take about a week to complete.

Guenter kept rather tight control on the guys who worked in the gantry area; those that were under his supervision. He would have considered it treason if any of the guys had talked with me or had any contact with me while I was in his domain. So, I didn't jeopardize their position there. The guys would talk to me in the elevator, or if we ran into each other elsewhere, but not up in the capsule area. I understood the situation and just did my job. Now and then one of the guys would look over at me and wink, but that was the extent of it. The only people I ever talked to while up on top of that gantry were the astronauts and Sam Beddingfield. Sam didn't report to Guenter and he didn't give a darn about a woman being up there. Of course, Guenter didn't care who knew he was trying to give me a difficult time. He thought that was so cute and clever of him. All I was trying to do was my job, but he couldn't stand having a female in his space. I never understood that since I certainly wasn't competing with him for his job.

I can't recall ever seeing any disagreement between one of the guys and Guenter. I think that if there were any problems like that they would have taken it somewhere private. However, he didn't mind letting lesser mortals know how he felt about certain things. I guess it was all a matter of rank and status with him. Peons didn't rate a private dressing down.

I painted Guenter one day. He was in his usual dictator-style mode and came over and stood very near me. I'd had enough and just reached over and painted a big red 'X' on his white smock. It was just plain luck that I was painting with red paint and it sure looked neat on that white coat. Needless to say, he wasn't happy with me. You'd have thought I'd painted his eyes red; they turned that color, he was so angry. I don't regret doing that. He was another of the original male chauvinists who felt a woman had no place up on top of the gantry. He and my boss really made a pair.

It seems this unreasonable hostility on the part of Guenter Wendt did not extend to another Chrysler graphic artist who temporarily invaded his work zone. A U.S. Army veteran named Howard ("Tony") Perez joined Chrysler's missile division in Huntsville, Alabama, and in late 1961 was sent to Cape Canaveral to work with NASA in the arts department with Cece Bibby. Whilst there, he painted portraits of all seven Mercury astronauts for the agency. Once Cece had finished her work on the *Friendship 7* logo, Perez was asked to paint the American flag on Glenn's capsule. There is no record of any conflict with Guenter Wendt (one suspects because he was a male artist) but Cece said that because of his work on painting the flag on the hull of *Friendship 7*, the arts department used to tease Perez by referring to him as the "Betsy Ross of NASA."

Pad leader Guenter Wendt, who did not appreciate Cece being in his work space. (Photo: NASA)

Howard ("Tony") Perez painting the U.S. flag on Glenn's spacecraft. (Photo: NASA)

MEN'S WORLDS AND MECHANICAL MISCHIEF

As Lawrence McGlynn observed in his interview article *Breaking Through the Glass Gantry*, "Bibby was a rarity, she owned and could repair her own sports car. She had a two-door AC Ace. The car was a British-made racing machine that Shelby would later use as a model for his famous Shelby Cobra series of race cars. It was bright red with white racing strips." Cece picks up the story:

> Before I did John's *Friendship 7*, I didn't know any of the astronauts, although I certainly would recognize them when I saw them. One day it looked like it was going to rain and I had the top down on the Ace, so left the office and went out to the parking area to put the top up on the car. The car was parked in a front row of the lot, facing the street. Across the street was the driveway into the hangar area where the astronauts had their office and training area.
>
> I had just finished putting up the top on the car when a blue Air Force sedan was getting ready to turn into the astronaut drive. The car had to stop for oncoming traffic. In the car was the driver and the passenger was Scott Carpenter. Scott did a double-take. He saw a female out there putting the top up on that red sports car, the car with the big white racing stripes on it. I mean he really did a double-take. I never forgot that. Later, when I got to know Scott, I mentioned that incident and he laughed. He said he had seen the car before and of course had mentioned it to the other guys. They all thought it belonged to some guy, so he was really surprised to see a woman out there acting like it was her car – which it was.
>
> The fact that it was the only vehicle like it in the area – probably in the State of Florida for that matter – really registered with people. I can even remember what I was wearing that day, after all these years. It was that much of a startling event. It was like Scott's head was built on a swivel.
>
> Later, when I was working in Houston, I remember one day when I parked the car at a grocery store. When I came out of the store I found about ten guys of all ages standing around the car looking at it. I put the bag of groceries in the passenger seat and then went around to the driver's side. You should have heard the questions: Was it my car or my husband's? Shocking that a single female would have a car like that. Was I sure I knew how to use a stick shift with five speeds forward? Did I know about double shifting? About down shifting? Like being a female automatically made me an idiot.
>
> I needed some repairs on it one day and took it to A. J. Foyt's garage; he was still racing at that time. I thought his father was going to pass out when he saw a woman driving that Ace. A. J. kept telling his dad to calm down.
>
> When I was at the Cape I used to take the Ace on sports car club rallies. The guy I was engaged to, a U-2 pilot, always insisted on doing the driving. I had to do the navigating even though I knew the area better than he did. Oh, well.

As Cece revealed, the Mercury astronauts were great practical jokers who enjoyed playing games on people – which Wally famously called "gotchas" – and everyone was fair game. "This was true, even amongst their own group. If you worked with them you had to accept the fact that you were a target, and the best you could hope for is that you might get one step ahead of them for a very short period of time."

An AC Ace similar to the one owned by Cece Bibby, but without the racing strips. (Photo: www.canepa.com)

As she recalled, the first astronaut she actually met was Wally Schirra. "When I was told to report to the astronauts' office to handle the mission logo for John Glenn, I was stopped at the gate. I did not have the proper ID tag to get into the building. The guard called their office and they said they would send someone down to escort me. I was standing there talking to the guard when he said, 'You must really rate. Look who they sent down to meet you.' It was Wally Schirra. Wally came out with that big Wally Schirra grin on and he was wearing a sport shirt, chinos and loafers. He came over and introduced himself to me and signed me in with that lovely Schirra voice. Then he took me upstairs to meet John Glenn." Glenn took Bibby around the office and introduced her to Deke Slayton, Gordon Cooper and Scott Carpenter. "I guess Wally had kind of clued them in that this young lady was there as the artist. I don't know whether they had expected a male artist or not, but I had a good conversation with them. Gordo made the comment to me, 'Well, you are not what we expected.' I asked what had they expected, and Gordo said, 'You don't giggle when we talk to you.'"

One of Cece's favorite recollections from that time concerns Wally Schirra, Gordon Cooper, a carrot cake, and a broken tail light.

> When the manned spaceflight program was just getting underway, *Life* magazine struck a deal… that would give *Life* exclusive access to the astronauts. The articles would cover a wide range of topics concerning astronaut training, their families, hobbies, etc. As part of the deal, *Life* rented a house in Cocoa Beach for the astronauts. The house was located about a mile or so south of downtown Cocoa Beach, such as it was. As it happened, the house was right across the street from the house I lived in. The house *Life* rented was in the high-rent neighborhood, being right on the beach, while I lived in the low-rent section across the street.

Mercury astronaut Gordon Cooper met his match with Cece Bibby. (Photo: NASA)

One day, heading home from work, I stopped at the store to pick up some groceries. I pulled into my driveway and began unloading grocery bags when Wally Schirra pulled in behind me. Wally helped me carry the bags into the kitchen. While in the kitchen, Wally saw a covered cake dish on the counter and lifted the lid to see what kind of cake it was. He thought it was a spice cake, but I told him it was a carrot cake. At that time that was a new recipe going around and I had wanted to try out the recipe. I asked Wally if he would like to try it and he said he thought it would be interesting. So I handed him a dish and a knife and told him to cut a slice. I also gave him a glass of milk to go with it. He said he really liked the cake and couldn't believe there were carrots in it. I offered him another slice to take across the street with him, and he thought that was a good idea.

The next evening, I had been home about half an hour when there was a knock on my door. It was Gordon Cooper. He said that Wally had told him I had baked a really good spice cake and he wanted to know whether there was any left. I thought it was kind of interesting that Wally had told Gordo that it was a spice cake and not a carrot cake. I figured that Wally was setting Gordo up for some reason and I would go along with it. Gordo finished off a piece of cake and a glass of milk and said he'd really like to have another slice. He was a little more than halfway through the second slice when he asked what ingredients were in the cake. I told him the recipe called for cinnamon but that I had also added nutmeg and some ground cloves. I then said, 'Oh, and grated carrots, of course.'

The fork in Gordo's hand stopped in midair. He got the strangest look on his face; he kind of gulped, and said, 'Carrots? You put carrots in this cake?' I offered to show him the recipe, but he decided against that. By this time the fork was now back down on the dish. Gordo pushed the dish away from him and said that he really had to be getting across the street, that he had some really important stuff to take care of right away. I asked whether he wanted me to wrap up what was left of his cake, so he could take it home with him, but he said he thought he'd had enough. I could tell by the look on his face that this was true.

The next day I saw Wally over at the astronauts' office and he said that when Gordo had come in the previous evening he looked almost ill. He'd said to Wally, 'That wasn't a spice cake Cece baked. She put carrots in it. Carrots!' Wally said that he put a very shocked look on his face so Gordo wouldn't know that he had been set up.

Wally and I got a big kick out of that. Here was a guy who was unafraid of going into space, but couldn't bring himself to eat a strange, new type of cake.

Cece may have thought she'd pulled a fast one over on Gordo, but the inherent risk in taking part in the "gotchas" was that retribution was always at hand for the unwary. He got her back a few weeks later – or so he thought.

One evening, soon after the carrot cake episode, she left home to attend a sports car meeting, driving her candy-apple-red Ace roadster:

It was the only one like it in the county – perhaps in Florida – and it really attracted a lot of attention. That being the case, I was very careful to obey the speed limit. I was on the causeway which crossed the river from Cocoa Beach to Cocoa when I was pulled over by the Florida State Patrol. I could not figure out what the problem was because I knew I hadn't been speeding. It turned out that the right taillight was out and I was given a warning to get it fixed. I went on to the meeting over in Cocoa, figuring I would replace the bulb the next day.

The next afternoon I was in my garage with the door open, trying to replace the bulb in the taillight, and found that wasn't the problem. There was a short in the wiring somewhere. Gordo came ambling over and wanted to know what was going on with the car. I explained the situation and said I guessed I'd have to take time off work so I could take it to have the wiring fixed. Gordo told me he thought he could fix it without any problem and, sure enough, he did. It only took him about 20 minutes and the taillights were working again.

The next morning it was raining so I put the top up on the car and, as I drove off, I turned on the car lights. When I got to the parking lot out at the Cape, one of the guys I worked with parked beside me. He told me that there was a problem with my taillights, that I had permanent brake lights. He said he noticed this when I went through the main gate at the Cape, where you had to slow down or stop to show your ID. I asked him whether he was sure, because I had just had them fixed. He got into my car, turned on the lights and depressed the brake pedal. There was no change in the density of the lights. He was right. When the lights were on I had permanent brake lights.

My comment then was, 'That rat!' He wanted to know what I meant, so I told him about Gordo 'fixing' my tail light. Well, my friend thought that was quite funny and got a good chuckle out of it. Strange sense of humor men have. I called over to the astronauts' office and was told Gordo was out of the country. I then remembered that he was scheduled to go to one of the tracking sites for Scotty's flight… it didn't matter because Gordo was out of reach and I couldn't wring his neck.

When I got home that evening I called a friend who was a U-2 pilot and he came over and fixed the wiring in the car. When he asked me how it had gotten that way I told him Gordo had 'fixed' it. Well, he thought that was a funny joke and chuckled the whole time he was working on it. Another guy with a strange sense of humor! I made up my mind I wouldn't say anything to anyone else about the taillight situation and when Gordo returned I'd just play it dumb. I figured that would really frustrate him.

Gordo returned to the Cape a couple of weeks later and came over to the arts department. He chatted with some of the other artists and then came over to talk with me. I asked him how the tracking site duty had gone and we talked about where he'd been… but no mention was made about the lights. Gordo asked how things had been going and I told him everything was going great. He sure had a puzzled look on his face. He couldn't figure out why I wasn't reading him the riot act for what he had done to my car.

I ran into John Glenn a day or so later and he asked what had gone on with my car. He said Gordo asked him and Wally if I had said anything about what he had done to the car. They told him they knew nothing about it, which was true because I hadn't mentioned it to them. John said they pressed Gordo about what he had done, but Gordo just clammed up and said he just thought I'd had some sort of problem with the car. I told John what had happened and that I was determined not to say anything to Gordo about it. John said he wouldn't let on to Gordo that I had gotten one of the U-2 pilots to fix the wiring.

I don't know whether Gordo ever heard about the outcome because I never mentioned it to him. I felt sure that if he had heard he would've had some smug comment to make about the joke he played on me. As it was, I think he was really frustrated trying to figure out the conclusion to the story.

Did I finally put one over on Gordo in the long run? I don't know.

"NEKKID LADIES" AND "GOTCHAS"

In addition to painting the logos on three of the Mercury spacecraft, Cece Bibby was notoriously enticed into a little mischief, thereby creating what she later described as the humorous tale of the "nekkid lady." This was her story:

> First, you should remember that the Mercury astronauts were very competitive guys, even among themselves, and each of them wanted to have the first flight. It didn't matter that the first couple of flights would be sub-orbital; first was first, and that was part of the attraction. These guys had really fought to be named the first astronauts, although some people referred to them as astronaut candidates. Wally Schirra once said that they were actually only 'half-astronauts' until a space flight was made. So, after the first seven were picked they then fought to make the first flight. This led to a lot of good-natured competition and jockeying for position and it involved every aspect of their flights.
>
> When Al made his flight there was a stencil cut for the name *Freedom 7* and the name was sprayed onto the capsule. The same was true for Gus's *Liberty Bell 7*. I don't know who sprayed the names on the capsules. I do know that when John Glenn decided that he wanted his *Friendship 7* hand-painted on his capsule there was a good bit of joshing that went on about it. Al and Gus made comments that a stencil wasn't good enough for John; that he had to have his name hand painted by an artist. Gus told me later that he wished he'd have had an artist do his *Liberty Bell*. He said it really bugged him that someone else thought of it and he hadn't. Competitiveness.
>
> Also, behind the scenes, every now and then, John would give the other guys lectures about their behavior. He'd tell them they were role models and, as such, they should always keep that in mind. I don't know just how seriously the guys took these lectures, but I think they just kind of let it all go over their heads. However, Gus would sometimes refer to John as their 'Boy Scout.'
>
> One day, as I was leaving the astronaut's office, I met Gus on the stairs. He wanted to know how the paint job was going with the 'Boy Scout's' capsule. Then he said he thought what I really should do was paint 'naked ladies' on the capsule because it would really shake John up, since he was so 'straight arrow.' Can't you just see that, naked ladies painted on the outside of that capsule? What an incident that would have created.
>
> I told Gus I didn't think that would be such a good idea. As he headed up the stairs to the offices he made some crack about the fact that I was chicken. I told him that it was my job that would be in jeopardy and not his. They weren't going to fire an astronaut. An artist was another matter. He just chuckled.
>
> I did get the idea that I could possibly play a joke on John by using the periscope view. I can't recall exactly what the periscope measured, but it was probably about 5 to 6 inches in diameter. Covering the periscope view was a pad on the exterior of the glass/lens, and the pad would be removed just prior to the countdown. So, I painted a naked lady. Well, she was lying on some throw pillows that were very strategically placed. The caption on this was, 'It's just you and me, John Baby,

against the world.' I should explain that because of my last name some of the guys called me Cece Baby or Cece Bibby Baby. I'm not sure, but I think some people thought my last name was Baby and not Bibby.

I had the photo lab make a print of the drawing and gave it to a friend, Sam Beddingfield. An engineer on the pad, he would be able to put the lady into the periscope view and then remove her just prior to the beginning of the countdown.

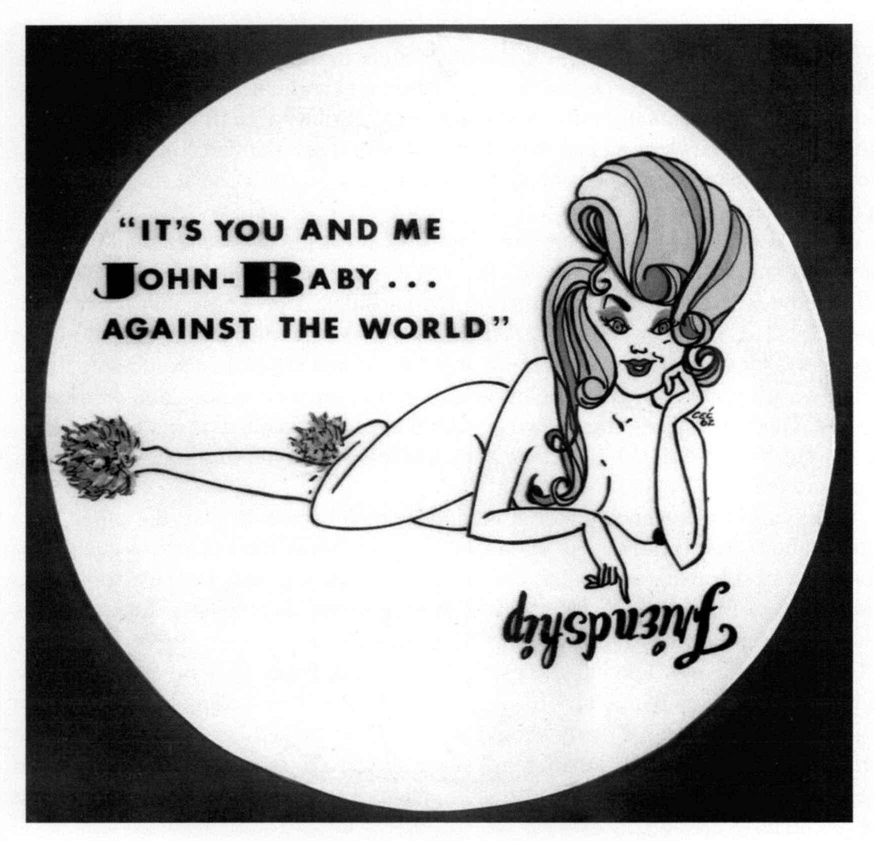

Cece's infamous "nekkid lady." (Photo: Cece Bibby)

As it happened, on the day the drawing was put in place, the flight was scrubbed. "When I came into work the next morning there was a note taped to the lamp on my drawing board. The note was from John Glenn telling me he had gotten a big kick out of the drawing. He also added that I shouldn't let anyone tell me that he was upset about the drawing." Based on that comment, Bibby figured there was a problem and there sure was.

"It is my understanding that [NASA manager] Rocco Petrone wanted to have me fired due to the naked lady painting. Petrone told my boss that I had upset John with the painting and could have caused the mission to fail. I showed John's note to my boss, but it didn't carry any weight with him. [But] I didn't get fired [because] John and Gus intervened and defended me. Gus even said he challenged me to the deed in the first place. My boss didn't dare fire me. He did try to get me banned from the pad, but that didn't work either. All the guys banded together and told management that they intended to have me design their insignias and paint them on their capsules. They really stepped up and protected me. I'll forever appreciate their help and friendship."

Bolstered by this strong support of the astronauts, Bibby put paint to paper again. "Just before John's next launch date, I did another lady for his periscope view. She wasn't what he expected. She was a rather frumpy old lady in a house dress. She had a mop in one hand and bucket in the other. The bucket had *Friendship* on it in the same script as his insignia on the capsule. The caption was, 'You were expecting maybe someone else, John Baby?'"

The follow-up gag. (Photo: Cece Bibby)

In a nice personal coincidence, John Glenn's Mercury mission was launched on 20 February 1962, which was Cece's birthday. Shortly thereafter, Scott Carpenter asked her to make a "nekkid lady" for him. "I told him I thought I could manage that, and I did."

According to Cece, the story behind Scott's "nekkid lady" went back to when Deke Slayton was replaced on his flight by Scott Carpenter instead of the MA-7 backup pilot, Wally Schirra. "Wally was upset because Scott got the flight to replace Deke. The way that it was supposed to be done was the backup pilot would take the flight. Instead they gave it to Scott. Wally's comment was that he hated to spend two months as backup for Scott because it meant listening to Scott play his guitar and sing *Yellow Bird*. That was the basis for my 'nekkid lady' and the caption on it. I really had to laugh at that. I had wanted to do a lady for Wally too, but thought it was pushing my luck after an incident that occurred prior to Wally's flight. It made me think twice about doing any more naked ladies."

Scott Carpenter's "nekkid lady." (Photo: Cece Bibby)

As Cece described that incident, NASA had published a manual about safety issues on launch pads. One item covered was the use of elevators, and a photo was included to show their correct use. In doing this, they had grouped six men in a pad elevator and photographed the proper way to ride in it. NASA distributed several hundred manuals only to find out that some enterprising artist had airbrushed out one of the men in the back row and substituted a very buxom blonde wearing nothing more than a smile. Once NASA found out they recalled all the manuals, but naturally enough they never managed to retrieve all of them.

"The finger of blame was pointed right at me," said Bibby. "I was in the clear though, because NASA had farmed out the contract on the manual and I had nothing to do with it. However, that cured me of painting naked ladies. I decided that I shouldn't push my luck with Wally's flight. I did a cartoon for Gordo, but gave it to him before his flight. It was Gordo with a bunch of ladies inside his capsule. They were lounging around on very colorful pillows and Gordo says something like 'Who brought the grapes?'"

As well as artistically interpreting hardware concepts presented by scientists and engineers, Cece's work in Chrysler's graphic arts department also diversified into the compilation of complex instruction manuals. One of the flight-connected jobs which Cece was asked to do was putting together Schirra's flight plan manual.

> The engineers, scientists, medical team – all had certain things they wanted Wally to do while he was in orbit. The public info office also had greetings they wanted him to give as he passed over certain countries. I had the job of entering all of that information on the flight plan that was to be installed in the capsule. It looked very similar to a roll of adding machine tape ... only wider. I really had to work to fit everything onto that flight plan. [It stated things like] when he'd have his blood pressure monitored, when he would perform certain technical jobs, when he'd send his best wishes to a certain geographical area ... like when he was flying over Africa or somewhere. Each time I thought I had the darned flight plan – or *thought* I had – Wally would give it back to me and tell me that he didn't want to do such and such at such and such a time. Maybe he didn't want to greet Mexico but would prefer to say 'hello' to people in Tasmania. That kind of thing.
>
> He did this to me so many times that I finally was close to blowing my top because it really involved a lot of changes. When I had all I could take of these changes I marched over to Hangar S and threw the darned thing on his desk and told him he was going to do it just as it was on the flight plan. NO MORE CHANGES. He started laughing, as did the others in the room, and he said, 'I was wondering how long it would take to get your goat.' I guess they had some kind of bet going about how long it'd take to make me get upset with him. I sat down and just stared at him. I finally had to laugh with him because of that darned laugh he has which is so infectious. I just shook my head at him.
>
> A few days later, I was in the cafeteria having lunch with some of the other artists when the NASA photographer sat down with us. He heaved a heavy sigh and said he wasn't sure he could even eat his lunch. He looked near tears. I asked him what the problem was and he said, 'That damned Wally Schirra – I'd like to strangle him.' It seems every time Bill [Taub], the photographer, was ready to shoot Wally's official photos Wally would turn his head to one side or the other. Wally would say, 'This isn't my best side. I want you to shoot my good side.' I told Bill that Wally was pulling his leg and told him about the flight plan mess. I asked Bill whether he had another sitting with Wally scheduled and he said he did, right after lunch, and he was afraid his stomach would be so upset that he would throw up. I kind of thought that regurgitating all over Wally was great retribution, but figured I should save Bill from

total embarrassment. So, I told Bill how he could fix that easily enough … all he had to do when Wally started in about his 'best side' was tell Wally that it was his opinion that Wally didn't have a 'best side' – that so far he hadn't even found a 'fair side,' let alone a 'good side.' And that's what Bill did.

He called me later and told me that when he did that Wally looked kind of stunned and then began laughing. He then asked, 'Who put you up to that?' Later, I went over to the hangar and poked my head in the office door. Wally looked up at me and I said, 'Well, I hear you had a time with the photographer. And that, Wally, is a gotcha!' He just roared that deep laugh of his. He said that he figured I would find a way to get even with him.

NASA photographer Bill Taub. (Photo: NASA)

AFTER NASA

Sigma 7 was the third and final logo designed and applied onto a Mercury spacecraft by Cece Bibby. Wally had told her he wanted to call his spacecraft *Sigma 7,* Sigma being a mathematical term meaning the "sum of." He felt that these flights were the sum of all the engineering, designing, testing, and construction which encompassed the entire Mercury program, and he wanted to honor all the people who had helped him make his flight.

In 1998, Schirra recorded an oral history interview with Roy Neal, describing why he liked the name. "Well, I think probably the best part of my Mercury mission was naming it *Sigma 7*. Naming it the sum of engineering effort. Not a fancy name like *Freedom* or *Faith* or *Aurora*. Not that I didn't appreciate those names. But I wanted to prove that it was a team of people working together to make this vehicle go. That's why I talk so wildly about knowing the engineers, how they were brothers and buddies. And all of them were! And that's what I saw as the ultimate on that mission, was that [it was] an engineering test flight, where we weren't going to look around for fireflies."

Cece begins work on painting the *Sigma 7* logo on Schirra's spacecraft. (Photo: NASA)

Cece drew up a couple of designs for Wally's review, and one featured the Greek symbol for Sigma. Wally liked this one a lot, and said he felt that Cece's design best represented his goal of making the MA-8 flight as perfect as possible. Which he later accomplished. "As with the other logos, most of it was lost during re-entry," Cece reflected.

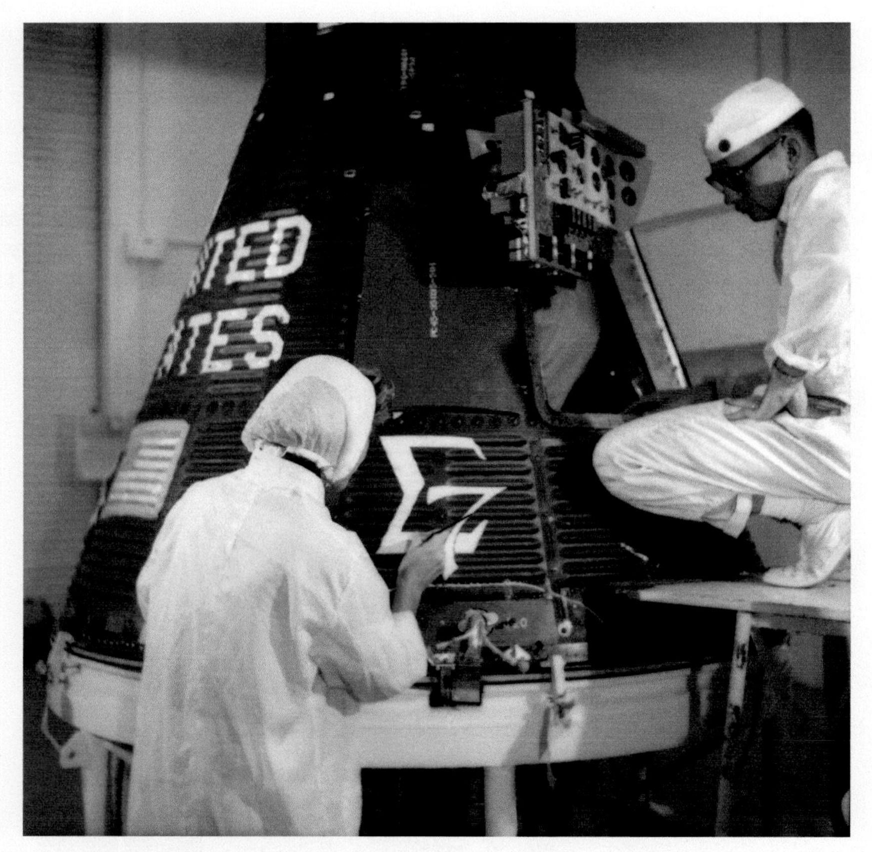

A pad worker watches on with interest as Cece's artwork nears completion. (Photo: NASA)

In 1962 Cece left NASA and relocated to California, so she was not around to design and paint a logo on Gordon Cooper's *Faith 7* capsule. However, as she told interviewer McGlynn, the thrill of the Cape and the launches lured her back. She returned shortly after Gordon Cooper's flight wrapped up Project Mercury, and in 1970 she married a naval officer. "As for Chrysler... I left them when I got married. My husband was a career navy officer and when he retired from the service he accepted a job in the offshore oil business. His job took us to Singapore for a couple of years, another two years in Bergen, Norway, and then a year in the Caribbean on the island of Antigua. So, I spent a lot of time roving the world."

Cece Bibby had experienced it all while working for NASA; she had also seen and heard some things in her time there that would have created massive headaches for the

Wally Schirra admires Cece's work. (Photo: NASA)

agency had they leaked out. But she was a trustworthy person, and kept confidences. "Yes, I do know some things that will never, ever see the light of day… at least not by me. There are some things that could hurt people and I don't want to do that. Besides, it's sort of water over the dam and what good would it do to bring some of that up now? I know that women threw themselves at all of the guys. At least what I saw at Cocoa Beach. And I'm sure that inflated some egos. Women flooded into the Cocoa Beach area from everywhere in hopes of latching onto the guys. I remember that a couple of hookers came up from Miami, figuring they would really make a killing financially in the Cocoa Beach area. They gave up after a short time and made the comment that they couldn't sell what was being given away. True story."

Posing with the completed Sigma 7 logo. (Photo: NASA)

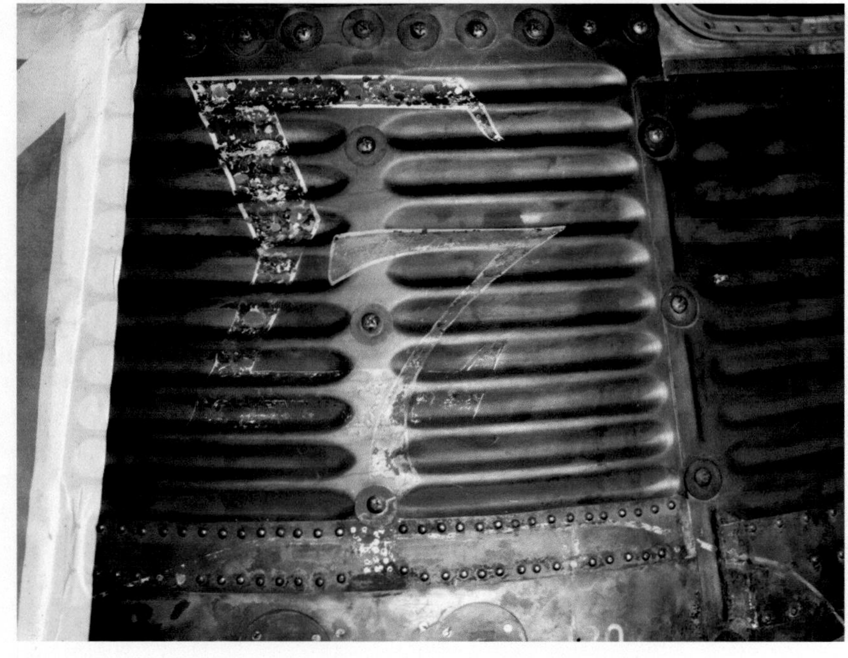

Post-flight, the *Sigma 7* logo can still be seen on the side of the heat-scarred Mercury spacecraft. (Photo: NASA)

Cece Bibby with long-time friend Lawrence ("Larry") McGlynn. (Photo: Larry McGlynn)

At a 2004 space show in Burbank, California, Cece was reunited with Wally Schirra, Gordon Cooper and Scott Carpenter, sporting a black eye after a fall. (Photo: Larry McGlynn)

Cecelia Bibby, living her later years in Blue Ridge, Georgia, died on 14 November 2012 at the age of 84, and is sadly missed by those who knew and loved her. On her tombstone are inscribed the poignant words, "You are not forgotten."*

---

* *The author wishes to thank spaceflight enthusiasts and historians Lawrence McGlynn and Bruce Moody, two great friends of the late Cece Bibby, for their generosity and support in supplying stories based on interviews and years of e-mail exchanges with the lady. For further information on Cece Bibby by Lawrence McGlynn see the following web pages:*
http://www.spaceartarchive.com/2013/10/cece-bibby-the-mercury-artist-.html
http://www.collectspace.com/news/news-080705a.html
http://www.space.com/news/cs_050909_bibby_bio.html

4

Sigma 7 flies

Following repairs in September to a leak in the Atlas rocket's fuel tank and a further five-day delay caused by a faulty valve in *Sigma 7*'s reaction control system, Schirra's flight was listed on the Atlantic Missile Range schedule for 3 October 1962. Launch time was set between 8:00 and 10:00 a.m. (EST) to provide for at least three hours of daylight search time in the six-orbit recovery area, located some 275 miles northeast of Midway Island in the Pacific Ocean.

As technicians worked to solve the faulty valve, NASA spokesman John ("Shorty") Powers hinted that if the six-orbit flight did not encounter any major problems, then a full-day flight, perhaps lasting 18 orbits, was a possibility for January or February with Schirra's backup Gordon Cooper the most likely pilot. NASA officials had agreed that a six-orbit flight would bridge the gap between the three orbits of John Glenn and Scott Carpenter and the full-day mission, and were anticipating some helpful results and data. As Powers explained, "A six-orbit mission would be beneficial to astronaut experience, overall operational training and spacecraft systems developed for the one-day mission to be performed later."

FINAL PREPARATIONS

In the lead-up to his flight, Schirra had been living in the astronauts' quarters in Hangar S at the Cape, undergoing numerous medical, psychological and other preliminary tests, all of which he passed with ease. In every respect he was primed and ready, both physically and mentally. On 21 September, despite some mild grumblings, he had been placed on a controlled low-residue diet by Air Force physician Dr. Howard Minners. In the interests of the mission, he went along with this discipline and would soon adjust to the low-residue food.

C. Burgess, *Sigma 7*, Springer Praxis Books, DOI 10.1007/978-3-319-27983-1_4

On 11 September NASA played host to President Kennedy at Cape Canaveral, with Schirra showing the launch facilities to the president. (Photo: NASA)

Meanwhile, three miles from Hangar S, technicians were working around the clock to have the spacecraft and the Atlas rocket fully prepared and checked for the planned lift-off the following Wednesday. The upcoming weather, always of critical importance around the recovery areas in the Pacific and Atlantic, was predicted to be reasonable for the next few days.

As recorded in their post-flight report, Richard Day and John Van Bockel, both of whom served in the Flight Crew Operations Division, observed that Schirra had fully participated in extensive training and spacecraft checkout activities in preparation for his flight.

Flight Director Chris Kraft and Wally Schirra run through the details of his flight plan. (Photo: NASA)

Final checks with pad leader Guenter Wendt. (Photo: NASA)

During the preflight preparation period [from July 11 on], the pilot was engaged in a diversity of activities often requiring considerable travel and resulting in a crowded schedule [and he] spent a large portion of his time in briefings and meetings concerning every aspect of the mission ... In addition, he completed such required training activities as recovery training, survival-pack exercises, acceleration refamiliarization on the centrifuge, and reviews of the celestial sphere at the Morehead Planetarium in Chapel Hill [North Carolina]. The pilot also logged 35 hours in the T-33, F-102 and F-106 aircraft during his preflight preparation period. Flights to maintain proficiency in high performance aircraft are considered an important phase of training because the pilot must maintain the ability to make rapid and accurate decisions under actual operational conditions.

Astronaut Schirra achieved a high level of skill on the procedures trainer in performing the turnaround and retrofire maneuvers. Use of the transparent gyro simulator and a very good understanding of the spacecraft control systems and their operation prepared him adequately for scheduled in-flight activities, such as control mode switching, flight maneuvering, drifting flight, and the gyro caging and uncaging procedures that cannot be properly simulated on the procedures trainer. Preparation in such areas as emergency procedures, mission anomalies, egress from the spacecraft, and recovery procedures was also satisfactorily accomplished.[1]

Everything, it seemed, was in readiness for the launch of the MA-8 mission, including the silvery space suit that Schirra had helped to develop for the Mercury missions (and would later assist in developing for the follow-on Gemini program).

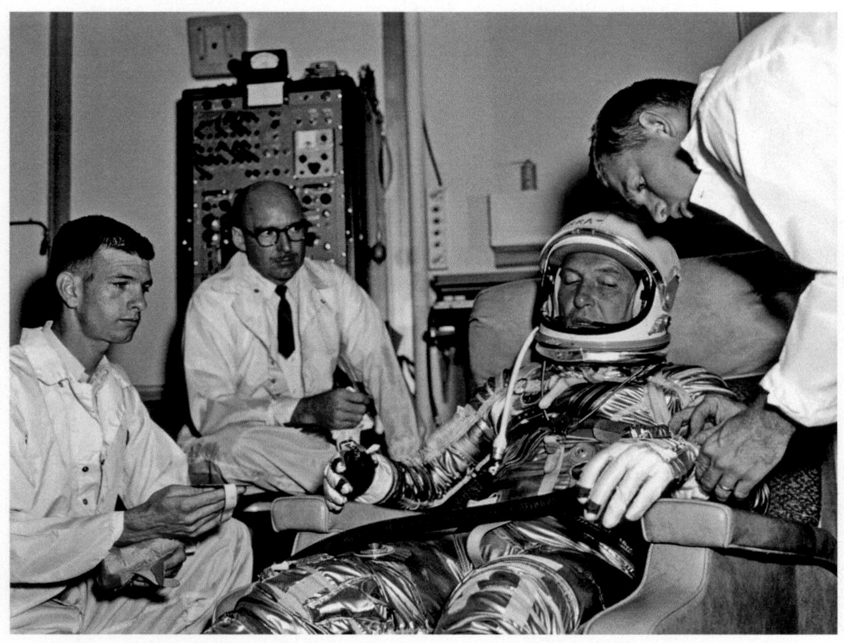

Schirra discusses the fitting of his space suit with (from left): suit technician Dick Sandrick, Dr. David Norris, and suit technician Al Rochford. (Photo: NASA)

With a little manual assistance, Rochford helps Schirra to don one of his suit gloves. (Photo: NASA)

For all but one of the Mercury launches, NASA's Joe Schmitt was the prime suit technician, with former Navy corpsman Alan Rochford serving as his backup. There would be an exception made on the MA-8 mission, however, when Rochford acted as prime suit technician for Wally Schirra. He had originally been assigned to fulfill this role on Scott Carpenter's earlier MA-7 mission, but a series of postponements meant that the anticipated launch date had just drawn too close to his wedding day, and he consequently asked Joe Schmitt and Scott Carpenter for permission to be relieved of this duty in order to make his way over to Philadelphia to get married. Naturally they both agreed. Scott Carpenter's mission aboard *Aurora 7* finally lifted off on 24 May 1962, and two days later Rochford married his bride Alice in front of their family and friends who had flown in from all over the country. Schmitt had meanwhile promised him he could take the next mission in line.

As expected, Rochford took on Schirra's MA-8 flight, working hand-in-hand with Schmitt. He admitted it was quite different acting as the prime suit technician, rather than the assisting backup role, but he said that Schmitt was a good teacher, always at hand to solve any problems. This gave him all the confidence he needed to carry out his tasks.

When asked how Schirra took to the numerous suit-ups that he had to endure during his MA-8 training, Rochford replied, "I suited Wally many times and he was always in a good mood. We chatted on all kinds of stuff – nothing serious, but [he was] always in a jovial mood."[2]

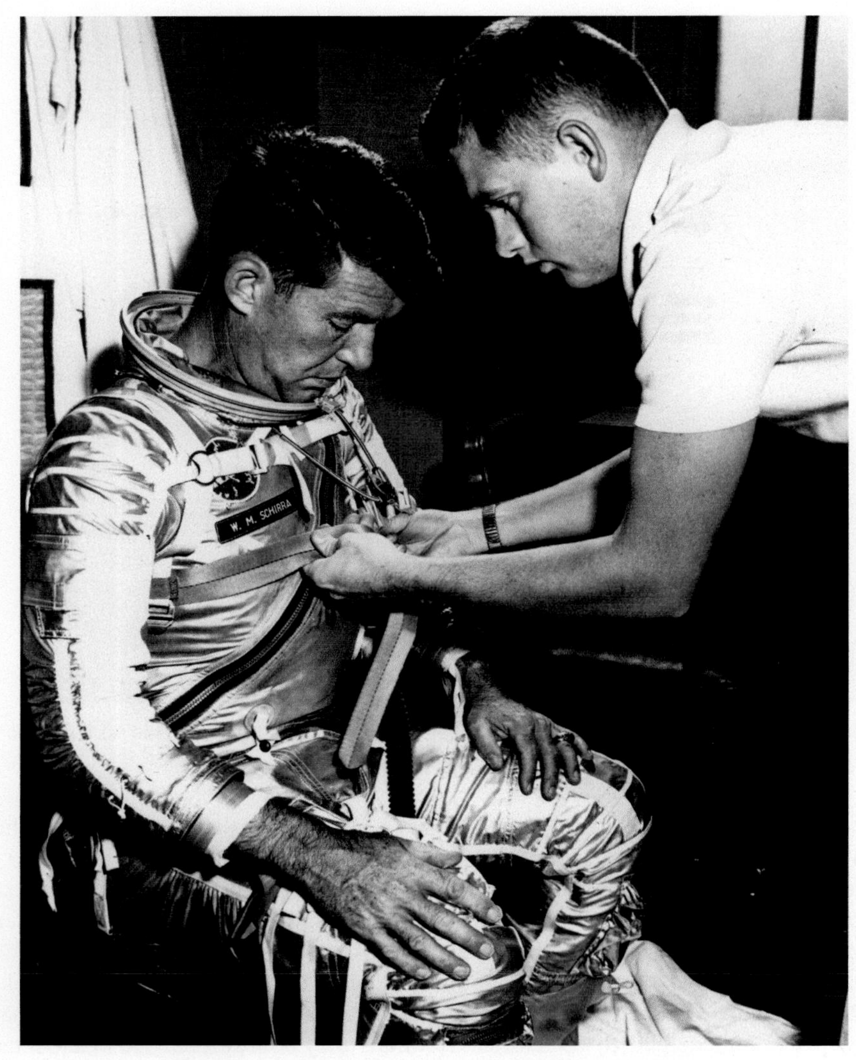

Al Rochford suiting up Schirra during training. (Photo: NASA)

Each Mercury astronaut had been supplied with three individually tailored pressure suits. One would be used for everyday heavy training such as water egress drills and would not be considered as the actual flight suit. The other two suits were used in simulator and capsule training. Close to launch day the astronaut would decide which of the two suits he would prefer to wear on the actual flight.

Schirra's weight fluctuated by as many as a few pounds during his astronaut years, and Rochford was asked if this presented any problems during the Mercury suit-ups. "Not during Mercury, but during an Apollo exercise I kidded him about closing his main entrance zipper. I knew he didn't want to talk about it, so [after that] I just kept my mouth shut."[3]

Schirra and Cooper engaged in a little pre-flight water-skiing. (Photo: NASA)

On Monday, 1 October, Schirra and his backup pilot Gordon Cooper completed physical tests at the nearby Patrick Air Force Base Hospital and passed with flying colors. Dr. Minners declared both to be in excellent shape. However, according to veteran space reporter Jules Bergman, that day could have ended badly for Schirra.

"Thirty-six hours before Wally Schirra was due to blast off, he was water-skiing on the Banana River, at Cocoa Beach. The occasion was a lawn party thrown by John Yardley, who then headed Project Mercury at Canaveral. Yardley was driving the speedboat with Schirra in tow – straight toward a sandbar in the river just behind Yardley's house. As we all watched transfixed, Yardley turned the boat, Schirra waved, and his skis hit the mud, somersaulting him into the air toward a tiny island in the river. I closed my eyes, knowing he had to be at least badly injured, if not killed – and there'd be hell to pay. But about two minutes later, Schirra swam into view unscathed: he'd gone clear over the island, landed in deep water and wasn't even scratched. Immediately after this, NASA instituted quarantine procedures, isolating astronauts for up to a week before a flight."[4]

That same day an unexpected weather problem loomed due to tropical storm Daisy, at that time some 300 miles north-east of Puerto Rico and moving in a north-westerly direction with stiff winds of up to 35 miles an hour. If Daisy were to continue in that direction, it could possibly bring stormy weather to the recovery areas where Schirra would land if he had to come down prematurely after making only two or three orbits. NASA spokesman "Shorty" Powers discussed the storm at a news conference and tried to be upbeat, "Everything is ready. All we can do is stand and watch the weather." He added that a decision would be made later in the day.

The following morning at 9:30, Maj. Gen. Ben I. Funk, Commander of the U.S. Air Force's Space Systems Division, carried out the Flight Safety Review to determine the readiness of Atlas 113-D. It was concluded that the Atlas was in suitable condition to support the MA-8 mission. NASA concurred.

In order to relax before a possible launch the next day, Schirra and Deke Slayton decided to do a little bluefish angling on a quiet beach near the launch gantries. One of these gantries housed a Thor Delta rocket, but the two men were only vaguely aware of its presence as they tried to catch the elusive bluefish, with Schirra finally landing one on the beach. All too soon, however, they realized they might have been in the wrong place at the wrong time. As Schirra later revealed, they were so intent on catching fish "that we were oblivious to activity at the pad. It wasn't until we heard a roar that we realized that the Thor Delta was lifting off. We were looking right up the tailpipe of its monster engine, and we knew right away that we were in the danger zone. Had there been an abort, it would have been a bad day for Mercury, with the chief astronaut and the pilot of MA-8 incinerated like the legendary rattlesnakes."[5]

Prior to retiring early for the night, Schirra called his parents, who were now living at Point Loma, San Diego. He was told that they would be watching the launch in the morning, gathered around their television set along with his sister Georgia Lou and her husband John Burhans, 13-year-old granddaughter Cindy, and their neighbors Mr. and Mrs. Gene Bowen. When Wally asked his mother what she had been doing all day, Florence replied that she had been busy in the kitchen baking things for the reporters who would be waiting outside their house. He was amazed.

"Mom, you don't have to feed them," he admonished.

"Well, I *want* to," was the curt reply.[6]

The Atlas rocket and *Sigma 7* stand ready for launch. (Photo: NASA)

ALL SYSTEMS "GO"

On launch day, Wednesday, 3 October, Rochford and Schmitt went through a well-rehearsed procedure. Both suit technicians were up before two o'clock getting ready for the big day, with Schmitt supervising, as Rochford recalled. "We'd get both of Wally's suits checked out; [make] sure there were no major leaks in the suit, because we had a leakage criteria that we had to stay under. So basically we'd lay out all his equipment, from urine bag to the underwear to the suit; all the different paraphernalia; the boots, the gloves, the helmets, and that sort of thing."[7]

Walt and Florence Schirra. (Photo: Schirra Family)

At 1:40 a.m. Schirra was awakened by his personal physician, Howard Minners. Backup pilot Gordon Cooper had earlier decided to sleep in a trailer outside Hangar S so that he wouldn't disturb Schirra when he left at 11:00 p.m. in order to run the final checks on the spacecraft.

Schirra then went through the routine of showering and shaving before having breakfast at 2:10 a.m. with Slayton, Bob Gilruth, Walt Williams, and Dr. Minners. Keeping it a secret for many years, he would later reveal that his breakfast included a freshly cooked one-and-a-half-pound piece of the bluefish he'd caught the previous day. Someone commented on the fact that the Russians had only sent up four cosmonauts to that time, but he would become the fifth American to venture into space. It was nice to know that – numerically at least – they were finally going to be ahead of their rivals in some way. He then began what he hoped would be his final physical examination and shortly after was pronounced fit.

Next, Schirra had electrocardiogram sensors attached to his body, to measure his temperature, blood pressure and heart action during the flight. As suggested by their physician Bill Douglas, each astronaut had four little tattooed spots to show the exact placement for each sensor.

At 3:25 a.m. Schirra made his way into the adjacent suit-up room, where Rochford and Schmitt were waiting for him, as well as Deke Slayton, with all the different parts of his suits prepared and ready for him to don. Despite the seriousness of the occasion, Wally still found time to play a pre-flight "gotcha" on nurse Dee O'Hara.

"Obviously, the key to the gotcha game is expectation, and Dee O'Hara played very well," he later recorded. "A piece of equipment we wore in those days without comfort facilities in space was a yellow plastic bag, otherwise known as a policeman's friend ... Dee was not there, but she had laid everything out. I looked, and to my dismay I saw this big black rubber thing. It was a urine bag that would fit a horse – about ten times the regular size. I put on a bathrobe and walked to Dee's office with this bag flopping between my legs, and she went into mock shock. 'Exactly the right fit,' she said."[8]

Suiting up for the final time; Al Rochford assists Wally Schirra on launch morning. (Photos: NASA)

Things then quickly returned to normal. "The photographers would be allowed to come in at a certain interval, once he had his underwear on, and he started getting into his suit," Alan Rochford mentioned during a NASA oral history interview. "Then still photographers would be allowed in, and the motion-picture guys. I remember we had Ed Thomas from RCA and Larry Summers from RCA come in. We had another NASA photographer work with us during those days, Bill Taub. As I recall, Bill was in the suit-up room. But anyway, they were allowed in for a short period of time while [I] did the suit-up. The Mercury suit was a fairly easy suit to get into, and once [I had Wally] zipped up and we put his helmet on and his gloves on and we got him on cooling, then we had a couch he could climb into. He would climb into the couch, and then we'd run [a manned air] leakage test on him ... and once that was completed, he would basically go out fully suited, connected to a battery-powered ventilator that was cooled with ice cubes."[9]

At 4:04 a.m. Schirra and his small entourage were ready to depart Hangar S and board the long transfer van that was waiting outside. Prior to closing his helmet and leaving, Schirra turned to some people who had gathered to see him off and said with his hallmark grin, "Well, I have nothing better to do today, so I guess I'll go out and take a trip around the world."[10]

Schirra leaves Hangar S on his way to the van that will transport him to the launch pad. Walking behind him are suit technicians Al Rochford and Joe Schmitt (in coveralls) and Dr. Howard Minners. (Photo: NASA)

Alvin Webb, a veteran space reporter assigned by the press pool to report on the walk-out from Hangar S, later observed that as Schirra made his way to the van, he seemed to be unusually relaxed and smiling, as compared to his predecessors at this time. Soon after, carrying his portable air conditioner, Schirra climbed aboard the van for the ride out to the flood-lit launch gantry 14.[11]

Al Rochford would travel on the transfer van along with Schirra, Schmitt and Dr. Minners. Schirra was fully suited with the visor on his helmet now closed, so he was breathing pure oxygen and purging nitrogen from his blood. "They spent much of the time discussing the flight," Rochford recalls, although Schirra says Howard Minners was doing most of the talking. In fact Schirra was so relaxed he motioned to Minners that he might close his eyes during the latter part of the 25 mph trip out to the pad and actually fell asleep. Minners woke him as they reached the launch pad and pulled up about 30 feet from the elevator which was to transport him up to the 11th level of the 160-foot-tall gantry.

When asked if the spare pressure suit was also in the van in case of a late problem, Rochford replied, "The backup suit stayed in the suit room, but we did take a spare helmet and some small stuff in the event something broke."[12]

STRAPPED IN AND SECURE

After a wait of 35 minutes, word came for Schirra to leave the transfer van and make his way to the gantry elevator. As he strode over to the elevator he stopped to shake hands with Byron McNabb, chief of the Atlas program for General Dynamics, who offered a short salutation. "On behalf of the crew of Launch Complex 14 we wish you a successful flight and a happy landing."[13] A grateful Schirra smiled, said "Thanks," and headed for the elevator. Gordon Cooper accompanied him in the elevator to the spacecraft level; he had spent much of the previous evening aboard *Sigma 7*, running a final check on all the switches and systems as the final countdown got under way. He would remain there until hatch closure, observing preparations for the shot.

With Cooper leading, Schirra steps down from the transfer van. (Photo: NASA)

With the gantry crew looking on, Schirra stops briefly to thank Byron McNabb and his team. (Photo: NASA)

During the elevator ride Cooper briefed Schirra on the latest weather reports and the state of the mission. Cooper also mentioned that the gauge for the manual-control fuel would show the tank on 94 percent full instead of 100 percent – Schirra was to ignore this

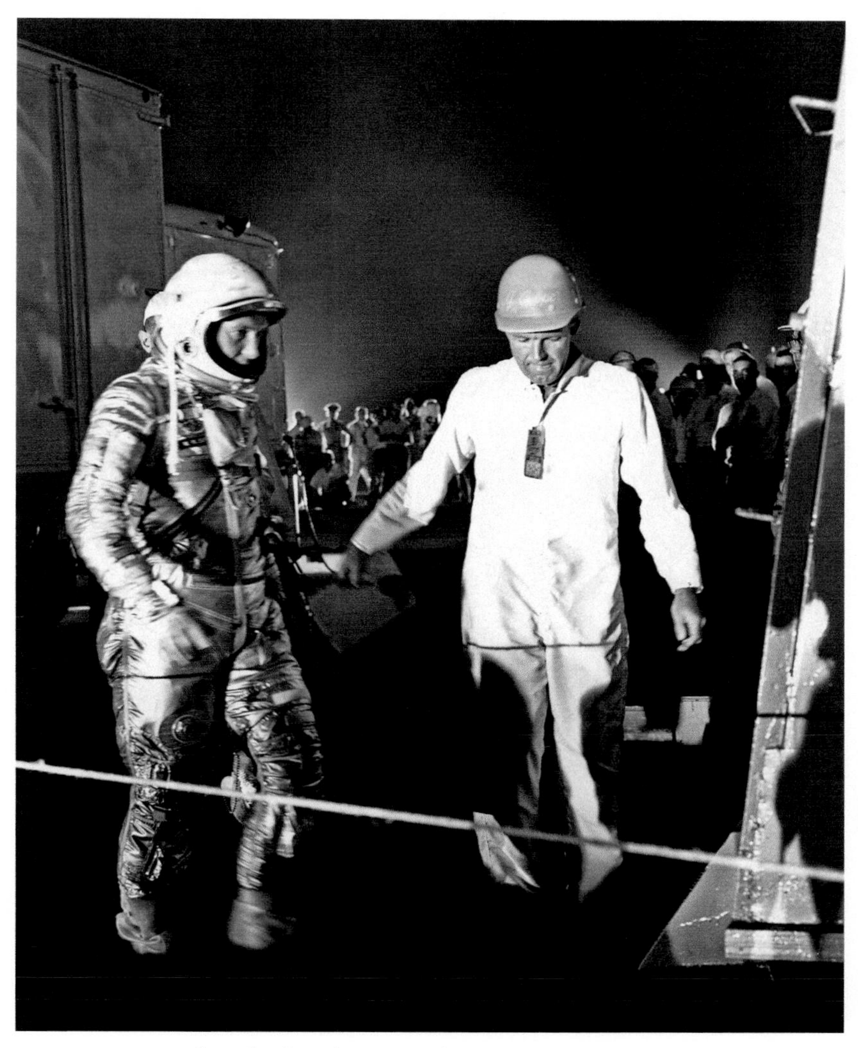

Stepping into the gantry elevator. (Photo: NASA)

reading because the fuel tank was quite full. "I just thought you'd be interested in that," he said with a laugh in his Oklahoma drawl. He knew his friend had a real fetish about fuel status.

As Schirra approached the spacecraft, pad crew leader Guenter Wendt recalls that Schirra was "joking and relaxed."

Schirra arrives at the 11th level of the gantry at 4:39 a.m., where *Sigma 7* awaits. (Photo: NASA)

Once Schirra was ready, Al Rochford helped him to doff his protective overshoes, and starting at 4:41 a.m., he and Cooper assisted Schirra to squeeze into the cramped confines of the spacecraft. "He'd have a handhold to grab hold of [in order] to swing himself into the couch, and then we'd hook him up to his oxygen system and close and lock his visor and attach all the restraints, the lap belt, the shoulder straps, that sort of thing, and wish him a good flight," Rochford said.

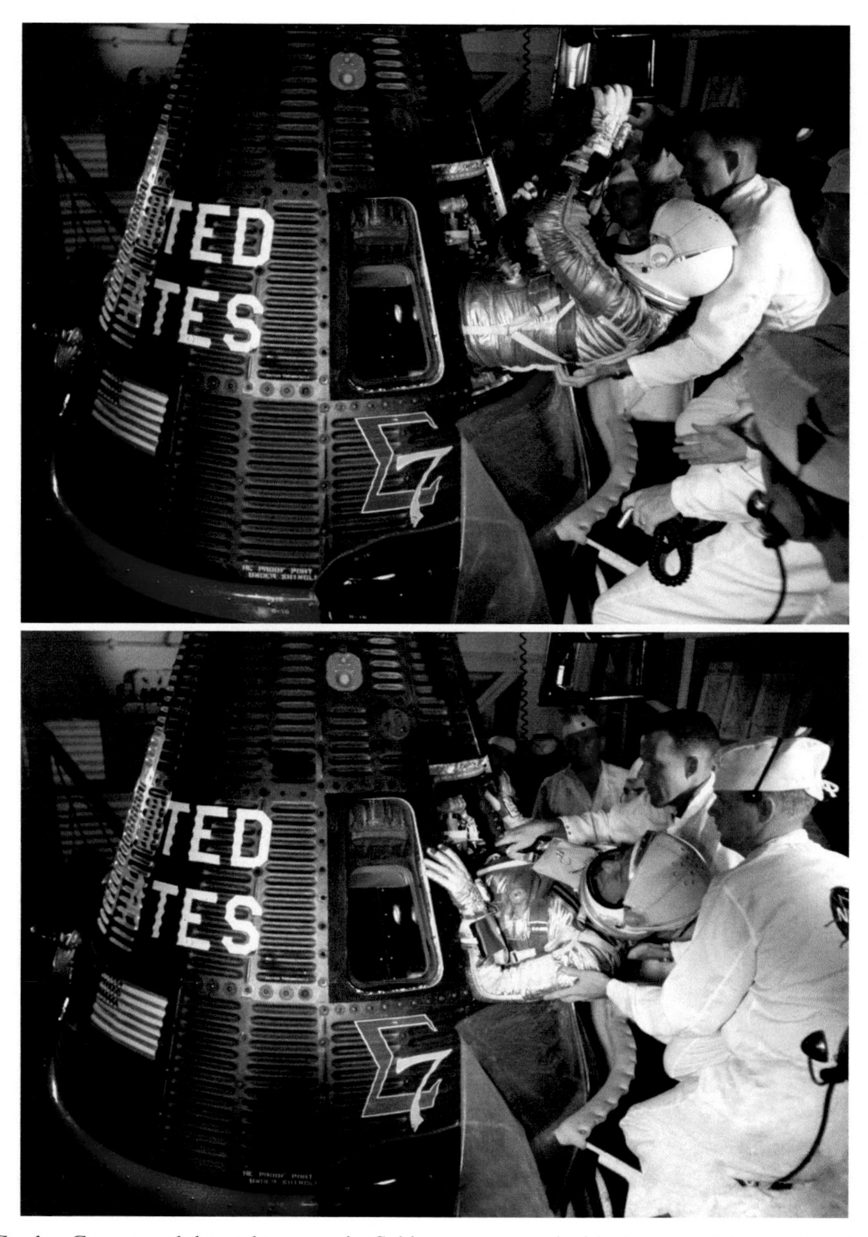

Gordon Cooper and the pad team assist Schirra to squeeze inside *Sigma 7*. (Photos: NASA)

Now nestled in his form-fitting couch and strapped in, Schirra began the routine of running his hands and eyes over all of the switches and instruments. As he expected, everything was in order.

Later, following the hatch closure, Rochford and others would head back down to the transfer van "and pull back to a fall-back area [to] wait for the rascal to go off."[14] According to Rochford, once the launch had taken place, the suit technicians would generally have very little to do. They would make their way back to Hangar S, gather up their equipment and head to Langley, Virginia to take on other duties. "There may be another astronaut getting ready for a training exercise, or something like that."[15]

Rochford was asked when his team would see the flown space suit again. After a flight, as he explained, the landing and recovery people would be responsible for an astronaut's pressure suit. "They'd bag it up, box it up, and send it back to us ... We would clean it up, perform post-flight tests on it [for] leakages, [conduct a] thorough examination to make sure it didn't have any tears or nicks or that sort of thing, and then when the crewman came back ... we'd get comments from him [on] how we could improve the suit."[16]

Meanwhile, as Schirra later reflected, "The camaraderie of everyone concerned with the flight preparations and equipment meant a great deal to me. For example, it was certainly a thrill while entering the spacecraft to see a dummy 'ignition key' on the control stick safety pin. This and other small gestures really helped to make me realize that there are many other people who were interested in what I was doing. We know this inherently, but these visible examples of it mean quite a bit. Here again, *Sigma* symbolizes the summation of the great efforts exerted by each and every man in the vast Mercury team."[17]

The dummy ignition key was not the only practical joke Schirra would discover. As he was rummaging around he opened a small personal storage compartment and found a steak sandwich wrapped in plastic, which brought forth another round of laughter. It was a gag initiated by the pad technicians, but NASA would later take a dim view of such levity. Of course, the sandwich was offloaded with gleeful thanks.

The spacecraft hatch was closed and bolted in place at 5:26 a.m., and unlike earlier Mercury flights would not be opened again until the mission had ended.

THE FINAL COUNTDOWN

An hour before the scheduled launch time, and once everyone had departed the launch complex, the gantry was rolled away from the launch pad and stationed in its parked position, leaving the gleaming Atlas poised and completely clear for lift-off. All that remained was to complete the countdown process, which was continuing on schedule.

At 6:15 a.m., some 50 minutes before the scheduled launch time, there was a brief delay when a 15-minute hold was called in the countdown. This resulted from a faulty motor in the radar tracking station in the Canary Islands, in the Atlantic west of Africa. But the problem was quickly rectified and the countdown resumed. It would prove to be the only snag in an otherwise perfect countdown.

Launch director Chris Kraft would tell *Life* reporter John Dille that, "This was not really a serious problem, and if Canary hadn't been able to fix it in another half hour, I would have let it go and gone on with the flight. The reason I hesitated at all was that Canary would give us a good fix right away on the exact orbital speed of the spacecraft. We wouldn't get such another good chance until Wally was over the tracking station in Australia. This is one of those marginal problems. It didn't have to be solved. We could live with it."[18]

Cooper shares some last words with Schirra before making his way over to the nearby blockhouse for the launch. (Photo: NASA)

The massive arc lights surrounding the launch pad were switched off as Schirra sat through the last pre-launch phase of the countdown. In his memoir *Schirra's Space*, he said he thought of many things as he waited for things to begin. He dwelt on his family, but also on the new house that they had built on a block of land at Clear Lake, south of Houston. Jo and the children had already moved in the previous month, but due to the intensity of his final training preparations he had not spent a single night in their new home.

"Of course I thought about how I'd handle the mission," he recalled. "I had made the point to NASA that my interest in public relations was zero. Nor did I intend to extol the wonders of space or portray the spectacle in vivid language. I was a pilot, an engineer, maybe a scientist. Sightseeing was low on my list of priorities. If it was poetry they expected, they should have sent a poet."[19]

Schirra had thought about shifting his mirror down in order to observe the engines lighting off and the coolant water deluge that would help dampen the severe vibration of lift-off. "But with the mirror down I would have covered the rate indicators, and I wanted to monitor those during boosted flight."[20]

Eight of the newly selected astronauts watch as the MA-8 countdown winds down. (Photo: NASA)

There were eight very eager observers at the Cape for the MA-8 launch. Just two weeks earlier, on 17 September, NASA had announced the nine names of its second group of astronauts. These men, whom the agency expected to pilot the Gemini and Apollo missions, had assembled (with the exception of Elliot See) to watch Schirra launch into space.

"Wally took particular delight in later years showing people a photo of the Group Two astronauts watching his launch," recalled space historian Francis French. "He'd ask you to look very closely, then point out that fellow astronaut Pete Conrad had his fingers crossed for good luck. He'd always give a huge, bellowing laugh when people spotted this."[21]

In the Launch Complex 14 blockhouse, Calvin ("Cal") Fowler, Test Conductor for General Dynamics, prepares to press the button that will launch the MA-8 mission. (Photo: NASA)

At 7:15 a.m., the almost perfect countdown reached zero. Inside *Sigma 7*, Schirra heard the two small vernier engines at the foot of the Atlas ignite and felt them begin thrusting. Then the sustainer and booster engines burst into lusty life. At that moment a torrent of orange flame exploded from beneath the mighty Atlas rocket which began straining against the hold-down clamps, ready to be unleashed and boost *Sigma 7* and Wally Schirra downrange on their way to orbit. "There is no doubt about when liftoff occurred," he later recorded in his post-flight summary. "If anything, I was somewhat surprised because it occurred earlier than I had anticipated."[22]

REFERENCES

1. Richard E. Day and John J. Van Bockel, report, "Pilot Performance," taken from *Results of the Third United States Manned Orbital Space Flight, October 3, 1962*, NASA SP-12, Manned Spacecraft Center, Houston, TX, December 1962
2. Alan Rochford email correspondence with Colin Burgess, 22 July 2015
3. *Ibid*

4. Jules Bergman, article "From Sputnik to Star Trek," *Flying* magazine, issue September 1977, pg. 318
5. Walter M. Schirra, Jr. with Richard N. Billings, *Schirra's Space*, Quinlan Press, Boston, MA, 1988
6. *The Daily News*, St. John (Newfoundland), "Mother and Father Watch Schirra on TV," issue 4 October 1962, Pg. 1
7. Alan Rochford email correspondence with Colin Burgess, 22 July 2015
8. Walter M. Schirra, Jr. with Richard N. Billings, *Schirra's Space*, Quinlan Press, Boston, MA, 1988
9. Alan Rochford interviewed by Summer Chick Bergen for NASA JSC Oral History program, Houston, TX, 15 September 1998
10. John Dille, "At the End of a Great Flight, Big Bull's-Eye," *Life* magazine, issue 12 October 1962, pg. 51
11. Alvin B. Webb, "Postlaunch Memorandum Report for MA-8," Part I; United Press International, "Hangar S Pool Copy," Oct. 3, 1962
12. Alan Rochford email correspondence with Colin Burgess, 22 July 2015
13. Bem Price, *Pittsburgh Post-Gazette*, "Schirra Lands Safely After 6 Perfect Orbits," issue 4 October 1962, pg. 6
14. Alan Rochford interviewed by Summer Chick Bergen for NASA JSC Oral History program, Houston, TX, 15 September 1998
15. Alan Rochford email correspondence with Colin Burgess, 22 July 2015
16. Alan Rochford interviewed by Summer Chick Bergen for NASA JSC Oral History program, Houston, TX, 15 September 1998
17. Walter Schirra, "Pilot's Flight Report," from *Results of the Third United States Manned Orbital Space Flight, October 3, 1962*, (NASA SP-12), Office of Scientific and Technical Information, NASA, Washington, DC, December 1962
18. John Dille, "At the End of a Great Flight, Big Bull's-Eye," *Life* magazine, issue 12 October 1962, pg. 51
19. Walter M. Schirra, Jr. with Richard N. Billings, *Schirra's Space*, Quinlan Press, Boston, MA, 1988, pg. 85
20. Walter M. Schirra, Jr., "Astronaut's Self-Debriefing," taken from *Postlaunch Memorandum Report for Mercury-Atlas No. 8 (MA-8),* NASA Manned Spacecraft Center, Houston, TX, 23 October 1962
21. Francis French email correspondence with Colin Burgess, 24 October 2015
22. Walter M. Schirra, Jr., "Pilot's Flight Report," from *Results of the Third U.S. Manned Orbital Space Flight, October 3, 1962*, NASA SP-12, Manned Spacecraft center, Houston, TX, December 1962

5

Six times around the world

As dawn broke on Wednesday, 3 October 1962, sports and space fans across the United States were faced with something of a quandary. Crowded around television and radio sets across the nation, people had tuned in to the expected early-morning launch of Wally Schirra aboard his *Sigma 7* spacecraft for the MA-8 mission. But that date also marked the opening World Series baseball game between the New York Yankees and the San Francisco Giants, so later in the day people would tune in their television sets to the baseball. During innings breaks, however, many would switch over to two of the three major TV networks in order to follow the real-life drama involving the orbital flight of America's fifth man into space.

Lift-off for the Mercury-Atlas occurred at 7:15:12 a.m. On the same date exactly twenty years earlier, the first successful launch of a V-2/A-4 rocket took place from Test Stand VII at Peenemünde in Germany, with Adolf Hitler and his Nazi cronies watching on. The missile eventually reached 52.5 miles altitude, thus becoming the first man-made object to reach space. In a speech at Peenemünde that day, Walter Dornberger, then head of Germany's infamous missile program, declared, "This third day of October, 1942, is the first of a new era in transportation; that of space travel." Two decades on, another rocket, Atlas LV-3B (113-D), was on a far more peaceful mission to carry spacecraft *Sigma 7* aloft on a steady path into the blue Florida skies. Astronaut Wally Schirra was finally on his way to becoming the ninth person to fly into space.

A NEAR ABORT AFTER LIFT-OFF

Things began to happen quickly. At the two-second point of his flight, Schirra transmitted the first of his mission reports, telling CapCom Deke Slayton, "I have the lift-off," in a calm and professional manner. "Clock has started. And she feels real nice."

Just six seconds later, as the mighty Atlas was powering into the heavens, belching a white contrail as it passed through the cool air, Slayton radioed a rather curious off-the-cuff comment that later had to be explained to puzzled news reporters.

"Wally; you got a pin for this flight?" Slayton enquired.

C. Burgess, *Sigma 7*, Springer Praxis Books, DOI 10.1007/978-3-319-27983-1_5

Pre-dawn at Launch Complex 14, and the Atlas rocket stands ready to blast *Sigma 7* into space. (Photo: NASA)

"Yeah, I got the pins on my office wall," Schirra replied, obviously understanding what the question meant.

The query harked back to a mental lapse Schirra had suffered several years earlier when he was an instructor in the U.S. Navy. Now in the first few moments of his flight into space this had come back to haunt him. As a Navy operations officer he had once solemnly admonished a group of young pilots to be sure to pull their landing gear pins before take-off. Shortly after issuing the warning, Schirra had taken off with one of the gear pins still installed. Embarrassed, he had to turn around and land again because he was unable to raise the secured gear. A few days prior to the MA-8 flight, an old Navy friend had jokingly sent him a pair of gear pins, which an amused Schirra had hung on his office wall.[1]

Lift-off for Wally Schirra and the MA-8 mission. (Photo: NASA)

At the 30 second mark into the flight, Schirra reported that all was going well. He was not to know that within seconds after lift-off telemetry signals emanating from the Atlas 113-D booster indicated there had been an unexpected clockwise roll. Primary and secondary sensors located inside the rocket rapidly assessed the potential threat to the vehicle, and transmitted data which indicated that the roll was only some 20 percent short of an abort condition.

"At ten seconds into the flight there was a problem I didn't know about then, but it came close to ending my fun. The clockwise roll rate of the Atlas was greater than planned, and it startled people in Mercury Control who were reading the instruments. My course was being plotted against an overlay grid called a harp, since it's shaped like the musical instrument. Green lines in the middle of the grid designate the safe zone, and on the outer limits the lines go from yellow to red. I was headed into the yellow area. If I had reached the red; there was a likelihood that the Atlas would impact on land, possibly in a populated area. In that situation the range safety officer would have had no choice but to abort the mission. He would have pushed a button to destroy the booster, and I would have had to depend on the escape tower. The tower would automatically pull the spacecraft away prior to the destruction of the booster, and I would have been carried by parachute to a landing in the Atlantic."[2]

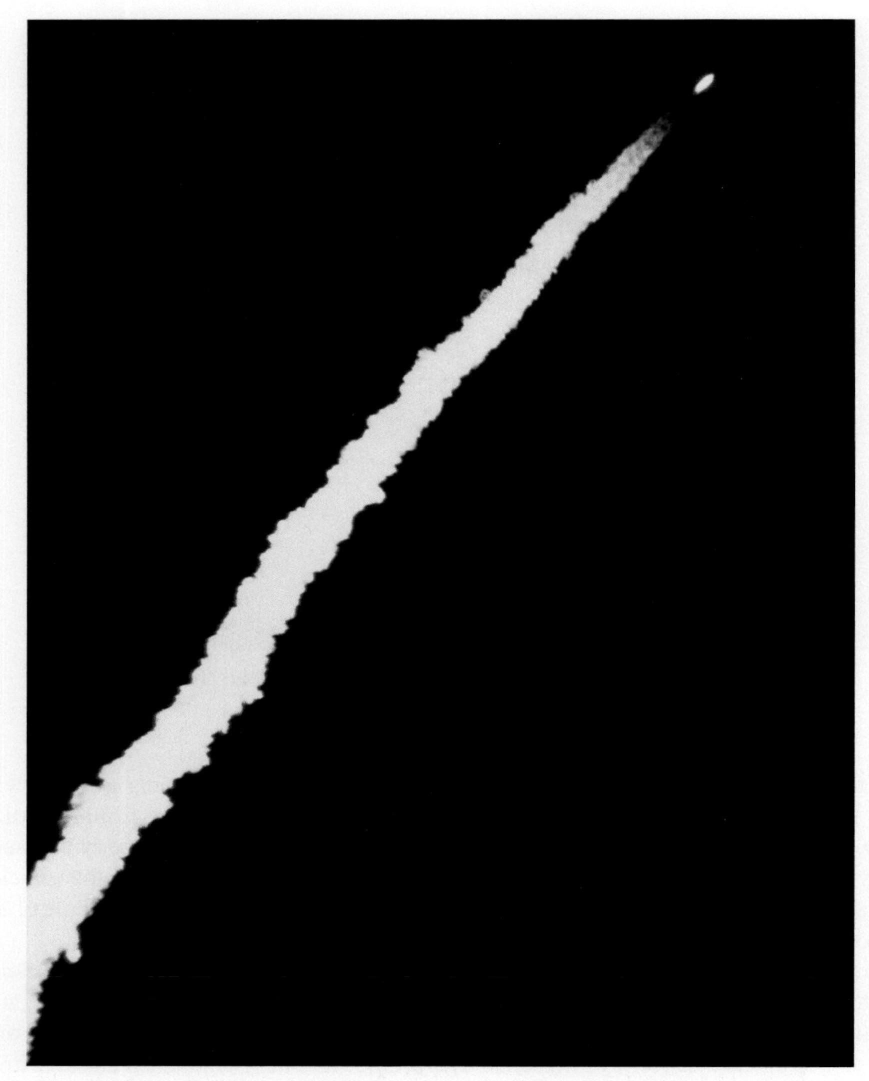

Cameras continued to track the ascent of the Atlas rocket as it soared into the skies. (Photo: NASA)

As a deeply concerned Mercury Control Center continued to monitor the situation they were relieved to note that the roll had discontinued, and the prospect of an abort situation rapidly diminished. "I was fortunate – the Atlas stayed within safe limits," Schirra later stated.[3] In fact, he didn't learn of this situation until after his flight. The fault was later identified as a slight misalignment of the main engines, but it was kept under control by the booster's vernier thrusters. "None of us ever did use the escape system in Mercury," he later commented. "It was not an exercise we cared to practice. All we know is that it would have been a rough ride with a high probability of injury."[4]

Mercury Control Center personnel track the flight of *Sigma 7*. (Photo: NASA)

Unaware of how close the mission had come to being aborted, Schirra continued to report to the Cape.

00:00:30 (Schirra): Okay. Fuel is okay. Oxygen is okay. All systems appear go, and she's getting noisy.
00:00:42 (Schirra): Not at all too noisy. Easy to talk through.

He continued to transmit readings on his instruments and cabin pressure, but to his puzzlement didn't receive any acknowledgement from mission CapCom Slayton, and found he was essentially talking to himself. He realized that his voice-activated radio microphone had been triggered by the thunderous noise briefly associated with flying through the point of maximum dynamic pressure (known as "max-Q") on the booster and spacecraft. This prevented any ground transmissions getting through to him. He quickly depressed the talk button, at which point he heard Slayton respond. The real concern for Schirra in this, was that activating the talk button required him to take his hand off the abort handle at a dangerous time in the ascent.

00:01:54 (Slayton): Roger. How do you read now, Wally?
00:01:55 (Schirra): I read you beautiful.
00:01:57 (Slayton): You had your transmitter keyed, and that's why we couldn't read.
00:02:00 (Schirra): I'll be darned. I'm push-to-talk now.
00:02:05 (Slayton): Stand by for staging.
00:02:07 (Schirra): I have a BECO. I could see the flash.

By this point in the ascent, the Atlas rocket was some 40 miles high and 45 miles away from the Cape and climbing at supersonic speed. At BECO (booster engine cut-off) the large outboard engines built by Rocketdyne ceased firing close to schedule – albeit just two seconds early. As staging took place the booster section with its two suddenly silent engines was explosively separated and fell away to an eventual impact with the ocean.

At BECO, the change in acceleration was quite obvious to Schirra. "I knew that the launch vehicle staged without having to wait for confirmation from the Cape CapCom, which, by the way, came in rapid order," Schirra commented in his post-flight report. "You can see the flashback of smoke from the booster engines as they part from the sustainer stage."[5]

The Atlas then continued to accelerate under the influence of the center-mounted sustainer engine and the two small side-mounted vernier engines, whose job was to guide and stabilize its flight. As the rocket's guidance system kept the engine firing through the center of gravity, the Atlas started to shimmy a little, but this vibration would soon pass as the guidance system pitched the Atlas forward so that it would go into orbit horizontally.

Just 25 seconds after BECO, the three small solid-fuel engines on the red escape tower – which was now no longer needed – ignited, and the jettisoned tower scooted away from *Sigma 7*, watched by Schirra.

00:02:27 (Schirra): Okay, I'm on push-to-talk, and the Sun is coming in the window now. Okay. There goes the tower.
00:02:35 (Slayton): Roger.
00:02:37 (Schirra): Auto retrojett off. This tower really is a sayonara.

As the tower blasted away from view, Schirra noticed that it left what seemed to be a thin film on the window. "I made a remark on the onboard tape about that during flight. The film that was left had colored splotches that were somewhat of an orange color. It could have been the orange paint from the base of the tower system, or [perhaps] some of the RTV-90 [silicon sealant]."[6] Throughout the entire flight he could see particles and filmy streaks on the outer window surface. That aside, everything seemed to be going smoothly. As he later recalled of the ascent:

The boosted flight itself was disappointingly short. Considerable training was conducted to prepare me for emergencies which might occur during powered flight. We so often practiced systems failures and aborts – either in the procedures trainer or by coordinating the trainer with the Mercury Control Center and Bermuda stations – that this practice made a very pronounced impression upon me … I never felt rushed, and all the events during launch were in order. I had more than the anticipated time available to me to make my systems checks. My scan pattern of the instrumentation panel was developed to where it was instinctive.

Since beginning this mission, I had become familiar with check-off points for various emergencies; for example, a no-BECO abort, a no-staging abort, and an abort at 3 minutes and 50 seconds after lift-off. It was a very pleasant feeling to check each of these off and put them behind me. This launch … was a successful, normal flight where I encountered many new experiences. I still believe that the amount of practice we had for the period prior to insertion is important, in that here the pilot must be prepared for reaction to an emergency, rather than thinking one out.[7]

ACHIEVING ORBIT

The next major event would occur a little over five minutes into the flight, at sustainer engine cut-off, or SECO. Schirra knew that if this came at the right moment, it was a sure sign the flight was progressing well. As that time approached, *Sigma 7* was close to 100 miles above the planet and being propelled along its set path at around 17,300 miles an hour. It was during this final phase of the ascent that Slayton famously tried (but failed) to pull a "gotcha" on the quick-thinking astronaut.

As Schirra later recalled: "At three minutes into my flight, when we were still in the Atlas boost phase, Deke Slayton came on the radio with a question I didn't quite expect. 'Hey Wally, are you a turtle?' Of course I knew the answer, but we were on live radio, and I wasn't ready for all the world to hear it. So I switched my mike to voice record, uttered my reply, clicked back to Deke and said, 'Rog.' After splashdown, several of us were in the admiral's quarters on the recovery ship, *Kearsarge*. Walt Williams, in his fast-chatter way of talking, demanded to know what my answer to Deke had been. I flipped on the voice recorder and there it was: 'Wally, are you a turtle?' 'You bet your sweet ass I am!'"[8]

It was a few moments of tension-easing levity, but Schirra quickly knuckled down to the more serious side of his flight. As the g forces built up once again, pressing him deeper into his couch, he could see the sky darkening to black outside.

At the five-minute mark Slayton radioed, "Stand by for SECO."

"It did seem that the buildup of acceleration during the sustainer period was rather slow," Schirra would later observe. "As I look back, the forces I experienced while being accelerated in boosted flight seemed to be much less than the later forces of re-entry. This comparison, I am sure, is best explained by the fact that you have a breathing point at BECO, in between the accelerations, while at re-entry there is a long continuous buildup of accelerations which are equally as exciting as those during boosted flight."[9]

SECO occurred at 5 minutes and 15 seconds into the flight. The sustainer engine had actually burned for about 10 seconds longer than intended, giving an extra 15 feet per second of velocity and inserting the spacecraft in a slightly higher orbit than was planned. Indeed, data received at the Mercury Control Center indicated that the added velocity had sent *Sigma 7* higher (176 miles) and faster (17,557 miles per hour) than any other astronaut had gone or would go during Project Mercury.[10]

Schirra glanced at his instrument panel, but he knew simply by sound and feel that the sustainer had shut down and the Atlas had been successfully jettisoned. Knowing from his previous training that there was no immediate hurry, he selected the auxiliary damping mode that would eliminate any lingering shuddering effect in the spacecraft caused by the blast of the posigrade rockets which had thrust *Sigma 7* away from the spent booster.

Having achieved orbit, Schirra selected the attitude stabilization and control system (ASCS) on the thrust selector switch, dropping into automatic pilot mode or fly-by-wire low. *Sigma 7* began to slowly turn around, to face the blunt end of the spacecraft in the direction of travel.

00:05:32 (Schirra): Yaw is answering very nicely. Roll answers nicely. She's turning around very nicely.
00:05:44 (Slayton): You have a go, seven orbit capability.
00:05:46 (Schirra): Say again. I like that kind.

"I resisted every impulse to look out of the window at this point, as I wanted to make the turnaround a fuel-minimum turnaround. The turnaround obviously in that case was done on the gyros. I discussed the turnaround as I was performing it, and got exactly what I wanted – approximately four degrees per second left yaw and had no trouble with any of the low thrusters at this time or ever after. I believe I got into retro attitude at about six minutes and fifty to fifty-five seconds. By this I mean in retroattitude to where I could have accepted retrofire if I needed it."[11]

A graphic artist's impression of *Sigma 7* orbiting the Earth. (Courtesy of artist Terry Leeds, San Diego Air & Space Museum)

In completing this slow turnaround with a deliberate intention of conserving fuel at every opportunity, Schirra had used only three-tenths of a pound out of the total loaded 59 pounds of hydrogen peroxide.

> **00:07:13** (Schirra): Roger. I just went into ASCS at about 7 minutes and 10 seconds. The sustainer is sitting very steady above me. I should say above the horizon. And I'm in chimp mode right now and she is flying beautifully.

Schirra, in his inimitable fashion, liked to call automatic control "chimp mode" as an insult. Once the turnaround had been completed, he could observe the sustainer stage through the spacecraft window. It was right where it had been predicted to be, and he was intrigued to see that the nozzle of the engine was pointing directly at *Sigma 7*. He reported that it was moving very slowly in relation to its insertion attitude, although it had somehow managed to make a 180-degree turnaround in the same time that he had completed his.

"I was also impressed with the fact that it was almost black in appearance, rather than the shiny silvery vehicle that astronauts [John] Glenn and [Scott] Carpenter had seen at this time and that I had observed on the launching pad. The white belly band of condensed moisture, the frost itself, was apparent to me. The sustainer followed the path that was predicted, and this knowledge helped to satisfy me that the attitude gyros and horizon scanners were operating properly. I did not see any crystalline material exhausting from the sustainer engine which Scott Carpenter had described. The sustainer, in retrospect, appeared slightly to the right of the predicted position which indicated a slight error to the left in my indicated attitude about the yaw axis."[12]

Now it was time for a momentous event; one Schirra had long been anticipating in preparing for his mission. He described this in a post-flight edition of *Life* magazine:

> At 10 minutes and 30 seconds after the launch, when I was about halfway across the Atlantic, Project Mercury made a real breakthrough in manned space flight. I flicked a switch that turned off all the automatic sequences – including any capability of the ground stations to control my retro-fire and bring me down. The capsule was all mine now.
>
> No astronaut before me had been permitted such freedom in orbit. The ground stations or the automatic sequences in the capsule had always kept fairly tight control of the situation. Now they were cut and the people on the ground were trusting me with the works. I had flown and tested airplanes, and they had my records on all that. But I had never flown the capsule before, much less under full pilot control.
>
> I was a pig-in-a-poke at this point and so, in a way, was the capsule. How well would the two of us get along? Could I lead and manage the capsule all by myself? It was a challenging, exciting moment. But not once – in the long planning and preparing, in the busy moments during the countdown or during the flight – did I doubt *Sigma 7* would do the job.[13]

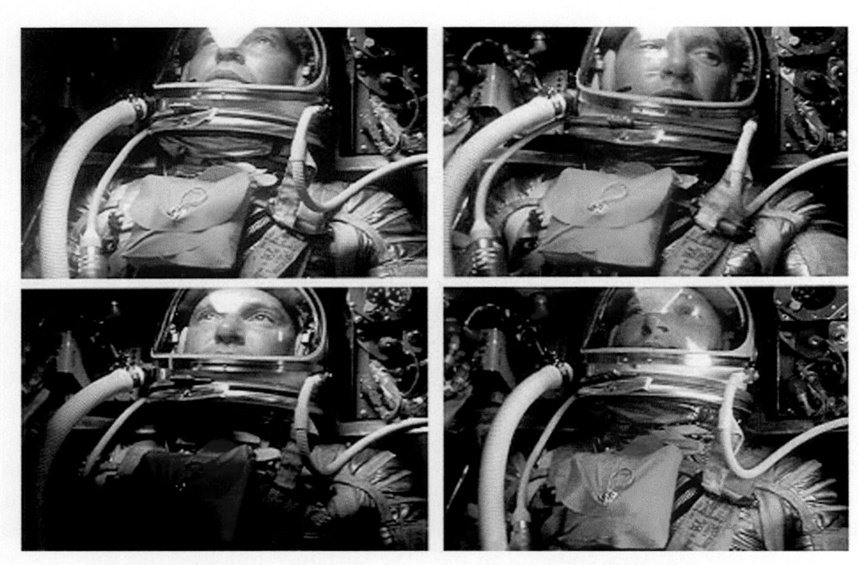

A sequence of photographs showing Wally Schirra in orbit. (Photos: NASA)

AN OVERHEATING SPACE SUIT

As Schirra approached the Canary Islands on his first orbit he decided to cease visually tracking the retreating booster and briefly switched to the manual-proportional control mode to run a test of the spacecraft by performing a pitching-up maneuver, which he found to be almost identical to those he had performed on the procedures trainer. But he judged this mechanical system to be "sloppy" and as he passed over Nigeria on the African continent he reverted to the ASCS autopilot control system. It was far better, he felt, to rely on the fly-by-wire system using low thrusters only.

"Africa itself, I didn't have much chance to assess as a viewing sight; I was much more interested in what was going on inside the vehicle. I did, of course, notice the desert terrain color of Africa; I found it very difficult not to notice it. The country itself was exactly as I anticipated from the chart I [had] in front of me at this time. Although at this time I was also well aware of the fact that we were working up to a suit system cooling problem. I decided then to devote my primary attention to solving this situation before I got yanked out of orbit. I was well aware of the fact that people were probably jumping up and down thinking 'here we go again' on another suit circuit crisis. And with proper concern I decided that I had better solve this one. We've had too many situations in this area and this one was real."[14]

Schirra was right to assume that the rising suit temperature might be cause for grave concern on the ground. In the Mercury Control Center, those who were monitoring the suit heat signal began to consult about a possible termination of the mission after just a single orbit. One controller closely monitoring the environmental control system was Frank Samonski, who had worked on the system with Schirra. He decided to discuss the astronaut's overheating problem with NASA Flight Surgeon Dr. Charles ("Chuck") Berry, chief of the Manned Spacecraft Center's Medical Operations Office. Berry was also monitoring

Schirra's condition and he ventured that the astronaut was still in good condition and the flight should continue, giving the astronaut a chance to work out the overheating problem. Flight Director Chris Kraft conferred with the two men, asking for their recommendation, and when they agreed that there was no immediate danger, he gave the go-ahead for the flight to continue into a second orbit.

As Frank Samonski explained, one of the larger problems they had tried to overcome was that of heat rejection using a heat exchanger and the evaporation of coolant water. "When water is exposed to a vacuum, it boils at a low temperature. Typically, without getting real technical, at about a tenth of a pound per square inch, water boils at about 35 degrees Fahrenheit, which makes it a very nice heat sink. But also, water freezes at 32 degrees. So you've got that very fine division between those two things. The water that was supplied to the heat exchangers, both the one in the suit circuit, and a smaller one in the cabin, was by [way of] two little water control valves, comfort control valves, they were called, CCVs, and they were like little metering valves. They supplied water, and the water flow went drip, drip. I mean, it was really, really slow."[15]

Having worked on the space suit's systems for some months, Schirra was well aware of the vagaries of temperature control, and he now set about resolving the overheating issue, as explained in NASA's seminal study, *This New Ocean: A History of Project Mercury*:

> When the temperature problem first appeared, the control knob setting was at Position 4. Prior to the flight, Schirra had established a procedure for just this situation. Rather than rushing to a high setting, he slowly advanced the knob by half a mark at a time, then waited about 10 minutes to evaluate the change. Had the valve been advanced too quickly, the heat exchanger might have frozen and reduced its effectiveness even more. By the time Position 7 was reached, Schirra was much cooler and felt sure that his temperature problem was nearing resolution, but for good measure he turned to Position 8. Shortly he became a little cool, and Samonski recommended that he return to Position 3.5. Schirra, thinking that some kind of analysis had been performed in the Mercury Control Center, complied. Immediately noting that the temperature was rising again, he quickly returned the setting to 7.5 and left it alone for a while.[16]

With 36 minutes and 57 seconds on the time elapsed clock, communications with the tracking station in Zanzibar were due to end. With the loss of signal approximately one minute away, Schirra was asked if he had anything else to report.

> **00:37:03** (Schirra): Nothing. I will keep the suit setting at this point until it gets a little hotter. If it does, I may have to go up another half notch at about 45 [minutes], before I get to Woomera.

Communications then became very garbled as *Sigma 7* flew out of the range of the Zanzibar tracking zone. During the few minutes before he flew into range of the next station, Schirra switched to voice transmit and record in order to keep an oral account on magnetic tape of his activities, well-being, and the spacecraft systems.

> **00:43:24**: I have switched to VOX transmit and record. I am satisfied that I can see yaw through the window on AMCS [Attitude Measurement Control System] without the use of the reticle by letting images come up from all sides. It's only a matter of a short period of time before objects show translation immediately.

00:44:07: The pitch scribe mark does indicate up a little bit and as a result matches the retroattitude, which at this time, is 30 degrees. I am now yawed right approximately 10 degrees, and it looks like I am tracking right down the line.
00:44:35: I am at 45 minutes. I am going to increase the suit setting knob just a small amount, about a quarter of a turn. I think we almost have control of the situation. I have set the suit knob at 6.25. The dome temperature at this time is 82 (degrees). Suit inlet is 76 (degrees).

At the 45-minute point of the flight communications had still not been restored, but Schirra continued to dictate his observations and reports.

00:45:10: I can definitely see a right roll at this time of about 5 degrees, and I noticed the periscope is dark, meaning we are coming into the dark side. I will attempt to look through the periscope for any observations. At this time, I can see nothing through the periscope for night observation, at least in this attitude. I am not even sure when I have low mag, other than the position of the lever. The window is cloudy. I have sunlight on it now and it has definitely been clouded over by the escape tower rocket, not to a great degree. I am seeing the so-called fireflies drift dramatically at this point. I tried a couple of knocks and they definitely have a relative velocity to the vehicle, but apparently are part of the same orbital system, I definitely see them as white objects.
00:48:19: With this much sunlight, I cannot see the stars at all. Sun is off to my left and I'm getting close to sunset at approximately 49 [minutes] is the scheduled time. That is just about right on. I'm approaching 49 and the cabin lights are on white. I am going to switch the cabin lights to red. And turn off that blasted lift-off correlation clock light.
00:49:13: Oh, I almost missed my first sunset trying to get the right cabin light off. It is rather rapid, as I was told it would be. I am not able to … There, I have got Arcturus right on the right side where it belongs.

Flying in the darkness of space after sunset. (Photo: NASA)

"GO" FOR A SECOND ORBIT

At approximately 51 minutes into the flight, Schirra was establishing communication with the Muchea tracking station in Western Australia. It was a place he knew well, as he had been an observer there for the flight of chimpanzee Enos the previous year, and he knew the technicians well. Unlike the four previous Mercury flights (MA-4 though to MA-7), the Senior Flight Controller (Capsule Communicator) duty did not fall to a fellow astronaut. For MA-8, this task had been assigned to Eugene Duret, a Canadian Avro engineer who had been hired by NASA.

00:52:15 (Duret): Has anyone asked you to get a drink of water, *Sigma Seven*?
00:52:18 (Schirra): Negative. I've tried not to get into that. If I can get the suit temperature down a little bit, I'll open the visor and get some water then.
00:52:25 (Duret): Roger. Understand.

Duret then advised Schirra about the forthcoming observation test using flares fired from Woomera in South Australia. He said the status of the test was okay at that time. There would be broken cloud over the test site and light rain, but no lightning had been reported, so they would be lighting the flares as planned. By then, *Sigma 7* was rapidly approaching the Woomera station.

00:56:29 (Schirra): Standing by for flare. Roll and yaw are holding. I see the flare on my left which is kind of wrong, I think. I think I saw a flash of lightning. Probably … that is lightning I am seeing, not the flare. I'm seeing more lightning. It is going to be hard to tell what I am seeing whether it is lightning or the flares.
00:57:09 (Woomera CapCom): *Sigma Seven* this is Woomera CapCom. Over.
00:57:12 (Schirra): Roger, Woomera. Go ahead.
00:57:15 (Woomera CapCom): Flare ignition will be in 1 minute 20 seconds.
00:57: 20 (Schirra): That's one reason why I can't see it, because I'm looking at lightning, obviously.
00:58:42 (Woomera CapCom): Ignition now.
00:58:43 (Schirra): Roger. I have lightning only. It looks like you are just about socked in. I'll stay here for a while and then come back up to ASCS shortly. I think I saw lightning right below me, but it couldn't have been the flare. It should burn steadily as I understand it.
00:59:03 (Woomera CapCom): Correct.
00:59:07 (Schirra): The lightning looks like a big blob, rather than a jagged streak we are used to seeing when Earthbound. Just looks like a big … almost like an anti-aircraft shot. A big blob of bright light, and then it fades out almost instantly. It definitely looks like you are overcast.

Leaving Australia behind, *Sigma 7* set out across the Pacific. As it approached the tracking station at Guaymas in Mexico, where CapCom Scott Carpenter would be his contact, Schirra reported that he was seeing his first sunrise.

01:24:40 (Schirra): I'm starting to see the sunrise in the periscope. First light in the periscope during this particular orbit as a result of the night side. It is obvious that the periscope has no function whatsoever in retroattitude on the night side. The sunrise is coming in rather rapidly through the periscope. I do have the lighted objects that John [Glenn] mentioned, and I can create some by knocking them off. I definitely have a sensation of their being a field and varying in size from small to bright.

The "objects" to which Schirra was referring here had caused quite a stir on John Glenn's MA-6 flight, when he reported tiny, bright particles drifting around near his *Friendship 7* spacecraft. At the time, he said they looked like "fireflies," and likened the experience to being surrounded by a field of floating stars. Totally mystified (as were ground controllers, who feared some sort of malfunction with the spacecraft), he banged on the walls of his spacecraft to see whether this would stir them up, which it did to an extent. Scott Carpenter later cleared up the mystery during his orbital flight aboard *Aurora 7*. He described the particles as resembling snowflakes or "frostflies," and this was actually quite close to the solution. They turned out to be nothing more than frozen condensation on the exterior of the spacecraft that became detached as it flew through areas of fluctuating temperatures.

01:25:00 (Schirra): The periscope itself is blinding me. I'll have to put the chart on it, so I can see out the window. I am in condition for retro at any time, so I have nothing else to do but look out this window. Assuming that the suit circuit is satisfactory. That chart helps no end to cover up that blasted periscope. Quite a large field of these objects. Definitely is confirmed that you can knock them off the hatch, as Scotty said. And they stream off at … Definitely there is no problem in judging that they are going away from the capsule, at a different rate than you are. They are definitely going slower, in velocity, than the capsule itself. One rap and you can see them sliding aft. They are too small an object for photography. I would not even attempt to take a picture of them.

Schirra then received a time check (which showed that he was one second ahead of ground elapsed time) and a welcome confirmation from Carpenter at Guaymas that he was cleared to continue into a second orbit, as the suit circuit problem seemed to have been resolved to everyone's satisfaction.

01:30:18 (Carpenter): One second fast and looks like you're good for another one, Wally.

After providing a status check on his condition and the spacecraft's systems and environment, Schirra reported further on what he believed to be the source of Glenn's mysterious "fireflies" that had garnered so much attention during the MA-6 flight and then caused Carpenter to waste so much time on his MA-7 mission attempting to pin down their source and photograph them through his window.

01:30:24 (Schirra): Okay. And I saw some of John's friends up here, I'm afraid to say; although I knocked them off the way you did it. Ha! Ha!
01:30:33 (Carpenter): Roger. Interested in your report.

01:30:44 (Schirra): Basically, what I saw was the firefly color that John saw, which I could create at other times in white color. I'm definitely convinced it's capsule ... a capsule derivative and once in a while, even now, I see one go by.
01:30:59 (Carpenter): Roger. That's good to hear.
01:31:02 (Schirra): I'm getting a very good yaw check with the yaw reticle in the ASCS mode. Having no trouble with that at all.

While appreciating that his astronaut colleague was back to reporting on the status of *Sigma 7*'s systems, Carpenter remained intrigued by the particles and wanted to pursue the subject. Schirra, though, had already lost interest in what was going on outside his window and wanted to move on.

00:31:12 (Carpenter): Wally, are the particles luminous or reflecting?
00:31:16 (Schirra): Scott, I think they are reflecting. I'm going to go ahead now, Scott, and do some yaw check as long as I've got some good terrain to look at and leave the particles alone for a while.

As Schirra explained post-flight in regard to the particles, "Water vented from the heat exchanger went overboard in the form of water vapor – not steam but molecular water. When it froze in the coldness of space, it crystallized, forming floating particles. The phenomenon was reported by both John Glenn and Scott Carpenter. Scott called them frost-flies. John saw them as fireflies, because of the prism effect when the Sun hits them, creating a color display."[17]

According to his later pilot's flight report, Schirra was very pleased with the way in which the flight was progressing, and was sticking rigidly to his mantra of completing a vehicle and systems test flight, rather than a mission overloaded with science and time-consuming experiments. He performed reference checks for the yaw attitude of *Sigma 7*, which was its left and right orientation, and these proved quite satisfactory.

"I do not believe I need to discuss the weather, the Sun, or the stars. It seems more appropriate to discuss the events within the spacecraft. Each network station got as much information as I had available to give them. Once we had solved the suit circuit problem and I'd begun to feel cool, I knew we were in a 'go' status and I had achieved my goal of using minimum fuel up to this point. I had stated long ago that I wanted to do some control maneuvers other than in fully automatic mode. I also had stated that I wanted to use the automatic mode when I did not need to employ manual modes or when I was too busy to fly the spacecraft, since this is why we have an autopilot. Admittedly, we have taken a system that was designed to be completely automatic and then tried to build some versatility into it and give the pilot the capability of controlling the spacecraft before I got to the Canaries, a fact which I reported to the ground. From that time on, I merely wanted to make observations that seemed to have merit and to use the control system only during those periods when I had to re-establish the attitude within the limits required to drop back into the automatic mode."[18]

As he flew over Cape Canaveral prior to commencing his second orbit, Schirra gave a status update to CapCom Deke Slayton, indicating he was still having some problems with his suit circuit, but it was no longer a major issue. Slayton asked him if he was drinking water in order to keep up his fluid intake.

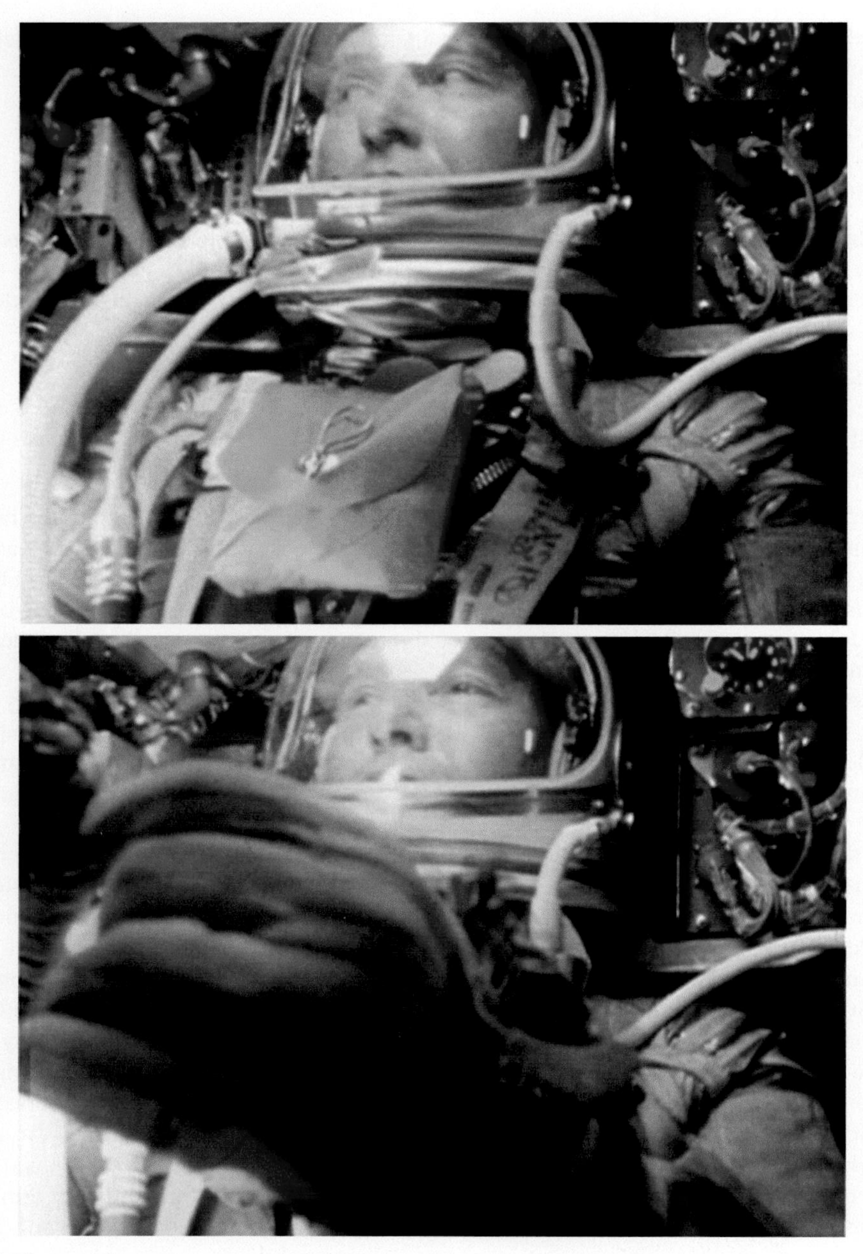

Schirra continues to carry out his flight plan aboard *Sigma 7*. (Photos: NASA)

01:37:36 (Schirra): No, I haven't. I've tried not to open the visor. I want to get the [suit] circuit going down. I think I might have a chance to take a quick one. I'll get ready for one.
01:39:14 (Slayton): How you reading now?
01:39:17 (Schirra): I read you fine. I just had some water and it does feel kinda good.

Although he did not take an abundance of photographs from orbit, Schirra did manage to capture some fine images using the modified Hasselblad camera. (Photos: NASA)

A TEXTBOOK FLIGHT

The flight continued to go well. As he passed over Woomera a second time, Schirra reported that his suit temperature had now decreased to 72 degrees Fahrenheit and so was under control. (For the remainder of the flight it would vary between 67 and 72 degrees.)

Passing over Cape Canaveral a second time he prepared for yet another exercise in fuel conservation. "I was to phase into drifting flight during the third orbit. I gave a systems status report to Slayton and I proceeded to cage the gyros, meaning to orient the gyroscopes to a precise attitude – and to cut off electrical power."[19]

As well, Schirra executed some simple experiments which he felt had future merit for astronauts, such as a psychometer experiment during drifting flight. This entailed closing his eyes and trying to touch certain designated dials on the control panel. Out of nine attempts, he only missed thrice, from which he concluded that weightlessness had not impaired his sense of distance and direction. However, there were still some surprises in store as he flew on in drifting flight:

> I was discouraged by the tremendous quantity of cloud coverage around the Earth and realized that it may always be a problem for certain space flight requirements. Africa, on the first and second passes, was ceiling and visibility unlimited (CAVU). The southwestern United States was also CAVU after I crossed over the ridge along the Baja California peninsula. I had a very good view, and I could easily determine yaw attitude by reference to the ground.
>
> When I re-established orbital attitude as I came over Muchea on the third pass, I was very pleased when I talked to the Muchea CapCom, and he and I agreed on yaw attitude exactly except for a possible four degree error in left yaw, which was also indicated by my instruments. The telemetered scanner readings were coincident with the spacecraft attitudes, and I'd just acquired these attitudes shortly prior to Muchea by using the Moon and the planet Venus, adjacent to it, for visual references. They actually showed up over the Indian Ocean [tracking] ship and were very easy to work with. They both lined up to give me a roll, pitch and yaw reference.[20]

As he passed over the tracking station based at Kauai, Hawaii on his third orbit, Schirra received some very welcome news from fellow astronaut and CapCom Gus Grissom.

> **04:24:43** (Grissom): Okay, fine. Cape feels you are in good shape Wally, and so I have some good news. They give you a go for six orbits.
> **04:24:48** (Schirra): Hallelujah!
> **04:24:51** (Grissom): They request you stay in retroattitude, and go ahead and prepare for retro like you would normally.
> **04:24:58** (Schirra): I understand.

Later, nearing the end of the fourth orbit, the MA-8 flight was still progressing well. In addition to monitoring the spacecraft systems and making lengthy, detailed reports on his condition and that of *Sigma 7* to the various ground tracking stations, Schirra had been busy using his photometer to measure a phenomenon known as "airglow," which accounts

The tracking station located on the island of Kauai, Hawaii. (Photo: NASA)

for much of the illumination in the night sky. As well, he had been taking a limited number of color photographs using the modified Hasselblad camera, and these would contribute to a catalog of space views of physiographic features such as folded mountains, fault zones, and volcanic fields. In addition, he took photographs of cloud patterns for comparison with those from other satellite programs. However, because a basic rule for the flight was to conserve fuel, photographs were taken when opportunity allowed.[21]

Back in touch with Gus Grissom in Kauai, Schirra was asked how things were going as he was about to complete his fourth orbit.

05:58:22 (Schirra): Oh, fine. I'm not bored up here. I just flew over Midway a while ago. Got a good look at that. I'm steaming up towards you all now, of course, north of you.
05:58:45 (Grissom): Go ahead with your report.

On his fourth orbit, Schirra photographed some cumulus clouds while flying over the West Atlantic Ocean. (Photo: NASA)

05:58:47 (Schirra): Okay, as you know, the control mode is set up for drifting. The mode selected is normal, auto, gyros caged. I've selected re-entry attitude. I'm in fly-by-wire low. The maneuver switch is off. I'll give you the fuel quantities and oxygen. Just to check yours against mine. I still have 89 [percent] auto and 90 [percent] manual.
05:59:20 (Grissom): Roger. Was that 89 – 90?
05:59:22 (Schirra): That's affirmative: 89 – 90.

His next point of contact seven minutes later was John Glenn at the Point Arguello station in California. Glenn asked if he had anything he would like to say because he was being broadcast live on TV for the next two minutes.

"Ha, ha," he responded. "I suppose an old song, *Drifting and Dreaming* would be apropos at this point, but at this point I don't have a chance to dream. I'm enjoying it too much."

For the MA-8 mission John Glenn was transmitting from the tracking station at Point Arguello, California. (Photo: NASA)

"BUENOS DÍAS, YOU-ALL"

Not everything went according to plan. Earlier, on his first orbit, Schirra had attempted to see the light from four flares from Woomera, each having an intensity of one million candlepower. A similar experiment took place at Durban, South Africa. However both experiments were unsuccessful because of extreme cloud cover. Schirra reported that, although cloud formations were prevalent around the entire ground track of the orbit, he was able to see lightning in a storm over Woomera and the lights of a city near Durban (later thought to be Port Elizabeth) while conducting the experiment. But passing over Durban on his sixth orbit he was at least able to see a xenon light of similar intensity for a period of about three minutes.[22]

As he was closing in on the eighth hour of his flight, now on his final orbit, Schirra was experiencing communication problems with Deke Slayton at Cape Canaveral. He was unexpectedly contacted by the tiny tracking station located at Quito, the capital of Ecuador, which was only supposed to contact him in the event of an emergency. The communicator ("whoever he was") agreed to pass on his message to the Cape, but then persisted in asking for a special favor at a time when Schirra was busy preparing for re-entry.

07:52:44 (Schirra): Hello, Quito. This is *Sigma Seven*. Can you relay to Cape that I read them loud and clear? Over.

Now on his final orbit, Schirra took this photo while flying over South America. (Photo: NASA)

07:52:50 (Quito): Yes, you are coming through fine. Any traffic you have, be glad to take it. Go ahead.
07:52:56 (Schirra): Everything here is all set. Would you relay to the Cape, I have everything under control. We are all set here.
07:53:02 (Quito): Very fine. Thank you very much. You don't have any word to pass on? Can you say anything in Spanish to the fellows down here?
07:53:12 (Schirra): I'm afraid I can't. Except I would like to come down and visit you. I'm enjoying a beautiful sight of the country.
07:53:19 (Quito): Certainly nice to hear that, but could you say just a few greetings to them? They would appreciate it so much. They want to put you on their radio down here.
07:53:27 (Schirra): I must send my greetings to the other people of our same area. The fact that we are two hemispheres joined is even proven today by our capability of flying over each other's country realizing that we are one and the same.
07:53:44 (Quito): Would you say "Buenos Días," or something like that back to them?

07:53:49 (Schirra): Right, all I can do on that now is say Buenos Días, you-all.
07:53:54 (Quito): Ha, ha. Thank you so much. I think they'll love that.

He later said, "I was furious. Here I was preparing for the crucial re-entry, and I was expected to utter some hogwash, some glowing statement about how I was proud to be an emissary of the United States, passing over Ecuador."[23]

Things then began to happen rapidly as Schirra ensured everything was prepared for the re-entry phase at the end of his sixth orbit. "It was time to get ready for the retrofire sequence. I switched to the fly-by-wire control mode while over Africa, and as I came up on the Pacific tracking ship, I positioned the spacecraft using celestial reference points. I reported my fuel supply to Shepard ... 78 percent in both the automatic and manual tanks. Then I switched to the automatic control mode and was satisfied that its high-energy thrusters were working well. I had the manual-proportional system as a backup. I was feeling fine, I assured Al."[24]

He was asked if he required any assistance in carrying out the pre-retrosequence checklist, and he responded with a very assured, "Negative."

HOMEWARD BOUND

Scott Carpenter's wasteful fuel expenditure during his three-orbit MA-7 mission had resulted in him starting for home with virtually all of his fuel depleted, which made it difficult to control his attitude during re-entry. The ground controllers were therefore highly impressed with Schirra's fuel conservation program over the six-orbit MA-8 flight.

As related in a post-flight report prepared by John Hodge, Gene Kranz and John Stonesifer, all from NASA's Flight Operations Division:

> Fuel management during the entire flight was exceptionally good. The spacecraft fuel tanks had been filled to capacity, and at the end of the first pass, the gauges for both the automatic and manual system fuel supply tanks indicated that there was approximately 98 percent remaining. In the majority of the cases where maneuvers were conducted over network stations, the fuel usage was so slight that it was difficult to determine if fuel was being consumed at all. The fuel usage in both the automatic and manual systems was significantly less than had been estimated based on the flight plan.[25]

In that same report, John Boynton and Lewis Fisher from NASA's Mercury Project Office also spoke enthusiastically about *Sigma 7*'s low fuel consumption:

> The usage rate of hydrogen peroxide control fuel was less than had been predicted for the MA-8 mission. The mission had been planned for minimum fuel usage, a philosophy which was incorporated into the schedule of inflight activities, and the astronaut adhered strictly to this flight plan. The result is especially satisfying when the fuel usage of the two previous flights is compared with that of the MA-8 mission, which was of much longer duration ... Although no additional fuel was included, a number of minor changes in equipment and flight procedures contributed to the increased fuel economy. The addition of a switch to disable the high thrusters when they were not needed permitted the pilot greater freedom in stick motion, since he then was not

required to restrain his hand movement within a fixed range to activate only the economical low thrusters. This switch, therefore, eliminated the possibility of inadvertently using the fuel-costly high thrusters during situations in which the pilot's attention might be distracted. The widening of the ASCS deadband was done in an effort to reduce the number and duration of control pulses per unit time and, therefore, the amount of total fuel consumed.

The primary technique to reduce fuel consumption, however, was the fact that the flight procedure included long periods of attitude-free drifting flight. During some of these periods, very small quantities of fuel were used at times to maintain spacecraft attitudes within the limits of the horizon scanners. When it was important to have the spacecraft at retroattitude for a possible mission abort, the ASCS orbit mode, which involves very small quantities of fuel, was utilized. Finally, the flight plan intentionally excluded control maneuvers which would have caused large quantities of fuel to be consumed.

It must be emphasized, however, that the previously mentioned factors were complementary to the pilot's discreet management of control system operations, for which he alone was responsible. Astronaut Schirra's discipline in using control fuel was the primary reason for the favorably low rate of expenditure.[26]

In preparing for re-entry, Schirra had completed the stowage of all items listed on the pre-retrofire checklist, and was ready well in advance of the event. It was time to come home after an incredibly successful and beneficial space flight.

I checked the high thrusters in fly-by-wire prior to retrosequence, and on the first demand for each high thruster in all three axes, they worked and reacted beautifully. It was a tremendous feeling to know that I had no problem with the high thrusters becoming cool. At the nominal retrosequence, the Pacific Ocean ship CapCom [Alan Shepard] gave a perfect countdown. Sequence and attitude lights actuated on time. I was sitting there ready to punch the retrosequence button. I did have the safety cover off the button and put it back on again. At the time of retrofire, the delay by a fraction of a second in firing the first rocket seemed agonizingly long. This time is probably the most critical of the flight, at least subsequent to insertion; and you know that these rockets *have* to work.

The rocket ignition was crisp, clean, and each one actuated with a definite sound. There was no doubt as to when each rocket was firing. The spacecraft did not seem to vary as much as half a degree in attitude during the period of retrofire. I was also cross-checking out the window and had plenty of visual cues in case things did go wrong with the automatic mode. I could see stars that did not even quiver. Because of these cross-checks, I was aware that the ASCS was working well throughout this period and did not require any manual control inputs.

Subsequent to the retrofire maneuver, I controlled the spacecraft with fly-by-wire. I had the retrojettison switch armed in time, and the retropackage subsequently jettisoned.[27]

Then, in scheduled compliance with a request from the engineers, he turned on the rate stabilization control system (RSCS). "The RSCS is a fuel guzzler," he would later write, "but I had plenty." He would also describe his fireball return through the heat of re-entry as "thrilling" and watched with a pilot's interest as the sky and Earth's surface began to brighten during his penetration of the atmosphere. *Sigma 7*, he said, was "as stable as an airplane."

It was hard not to feel elated.

REFERENCES

1. *The Daily News*, St. John's (Newfoundland), "Schirra Remembers Mental Lapse," issue 4 October 1962, pg. 1
2. Walter M. Schirra, Jr. with Richard N. Billings, *Schirra's Space*, Quinlan Press, Boston, MA, 1988, pp. 85–86
3. Francis French and Colin Burgess, *Into That Silent Sea: Trailblazers of the Space Era, 1961–1965*, University of Nebraska Press, Lincoln, NE, 2007
4. Walter M. Schirra, Jr. with Richard N. Billings, *Schirra's Space*, Quinlan Press, Boston, MA, 1988, pg. 86
5. Walter M. Schirra, Jr., "Pilot's Flight Report," from *Results of the Third U.S. Manned Orbital Space Flight, October 3, 1962*, NASA SP-12, Manned Spacecraft Center, Houston, TX, December 1962
6. Walter M. Schirra, Jr., "Astronaut's Self-Debriefing," taken from *Postlaunch Memorandum Report for Mercury-Atlas No. 8 (MA-8),* NASA Manned Spacecraft Center, Houston, TX, 23 October 1962
7. Walter M. Schirra, Jr., "Pilot's Flight Report," from *Results of the Third U.S. Manned Orbital Space Flight, October 3, 1962*, NASA SP-12, Manned Spacecraft center, Houston, TX, December 1962
8. Extract with permission from Wally Schirra website at: *http://www.wallyschirra.com*
9. Walter M. Schirra, Jr., "Pilot's Flight Report," from *Results of the Third U.S. Manned Orbital Space Flight, October 3, 1962*, NASA SP-12, Manned Spacecraft Center, Houston, TX, December 1962
10. Loyd S. Swenson Jr., James M. Grimwood and Charles C. Alexander, *This New Ocean: A History of Project Mercury.* NASA Special Publication SP-4201, Washington, DC, 1989, pg. 473
11. Walter M. Schirra, Jr., "Astronaut's Self-Debriefing," taken from *Postlaunch Memorandum Report for Mercury-Atlas No. 8 (MA-8),* NASA Manned Spacecraft Center, Houston, TX, 23 October 1962
12. Walter M. Schirra, Jr., "Pilot's Flight Report," from *Results of the Third U.S. Manned Orbital Space Flight, October 3, 1962*, NASA SP-12, Manned Spacecraft Center, Houston, TX, December 1962
13. Walter Schirra, "A real breakthrough – the capsule was all mine," *Life* magazine, Vol. 53, No. 17, October 26, 1962
14. Walter M. Schirra, Jr., "Astronaut's Self-Debriefing," taken from *Postlaunch Memorandum Report for Mercury-Atlas No. 8 (MA-8),* NASA Manned Spacecraft Center, Houston, TX, 23 October 1962
15. Frank H. Samonski, interviewed by Jennifer Ross-Nazal for NASA JSC Oral History program, Houston, TX, 30 December 2002

16. Loyd S. Swenson Jr., James M. Grimwood and Charles C. Alexander, *This New Ocean: A History of Project Mercury*. NASA Special Publication SP-4201, Washington, DC, 1989, pg. 475
17. Walter M. Schirra, Jr. with Richard N. Billings, *Schirra's Space*, Quinlan Press, Boston, MA, 1988, pg. 87
18. Walter M. Schirra, Jr., "Pilot's Flight Report," from *Results of the Third U.S. Manned Orbital Space Flight, October 3, 1962*, NASA SP-12, Manned Spacecraft Center, Houston, TX, December 1962
19. Walter M. Schirra, Jr. with Richard N. Billings, *Schirra's Space*, Quinlan Press, Boston, MA, 1988, pg. 87
20. Walter M. Schirra, Jr., "Pilot's Flight Report," from *Results of the Third U.S. Manned Orbital Space Flight, October 3, 1962*, NASA SP-12, Manned Spacecraft Center, Houston, TX, December 1962
21. John H. Boynton and Lewis R. Fisher, "Spacecraft and Launch-Vehicle Performance," from *Results of the Third U.S. Manned Orbital Space Flight, October 3, 1962*, NASA SP-12, Manned Spacecraft Center, Houston, TX, December 1962
22. *Ibid*
23. Walter M. Schirra, Jr. with Richard N. Billings, *Schirra's Space*, Quinlan Press, Boston, MA, 1988, pg. 89
24. *Ibid*
25. John D. Hodge, Eugene F. Kranz and John Stonesifer, "Mission Operations," from *Results of the Third U.S. Manned Orbital Space Flight, October 3, 1962*, NASA SP-12, Manned Spacecraft Center, Houston, TX, December 1962
26. John H. Boynton and Lewis R. Fisher, "Spacecraft and Launch-Vehicle Performance," from *Results of the Third U.S. Manned Orbital Space Flight, October 3, 1962*, NASA SP-12, Manned Spacecraft Center, Houston, TX, December 1962
27. Walter M. Schirra, Jr., "Pilot's Flight Report," from *Results of the Third U.S. Manned Orbital Space Flight, October 3, 1962*, NASA SP-12, Manned Spacecraft Center, Houston, TX, December 1962

6

"A sweet little bird"

After a period of modernization, repair, and training, the *Essex*-class attack aircraft carrier USS *Kearsarge* (CVS-33), known to her sailors as "Mighty Kay," departed Long Beach on 1 August 1962. A few days earlier, the carrier had been named the major unit in the Pacific Recovery Force for the MA-8 astronaut recovery operation. The commander of the Project Mercury Recovery Force was Rear Adm. Charles A. Buchanan (Commander, Hawaiian Sea Frontier). The designated Recovery Group Commander was Capt. Thomas S. King (Commander, Destroyer Flotilla Five). He would be on *Kearsarge* to control the operations of vessels within the immediate recovery area.

The role for the crew of *Kearsarge* was to proceed from Long Beach to San Diego, take on board aircraft and personnel from CVSG-53 (or Carrier Anti-Submarine Air Group 53) together with a detachment of VAW-11 fixed-wing aircraft, then arrive at Pearl Harbor on 7 August for the installation of special communications equipment. Finally, the carrier would take up an assigned station in the Pacific missile range and prepare for the MA-8 Mercury flight of Wally Schirra aboard his *Sigma 7* spacecraft, which was then expected on 18 September.[1]

USS *Kearsarge*, commissioned in March 1946, sailed under the command of Capt. Eugene Rankin, who had been accorded the honor of command of the mighty carrier just two months earlier.

RECOVERY TRAINING

Training aboard *Kearsarge* for the retrieval of *Sigma 7* and Wally Schirra started in August with its personnel attending recovery technique briefings conducted by Capt. Thomas King. The next step involved a "boilerplate" version of a Mercury capsule.

C. Burgess, *Sigma 7*, Springer Praxis Books, DOI 10.1007/978-3-319-27983-1_6

USS *Kearsarge* under way in September 1962. (Photo: U.S. Navy)

Initially, the unmanned dummy capsule was towed around Pearl Harbor using the carrier's motor whaleboat. The recovery concept was that the whaleboat crew would latch onto the capsule and tow it to the starboard side of *Kearsarge*, whereupon a hoist would lift it onto the carrier's deck. After rehearsing in the harbor's smooth seas, an attempt was made in open water. However, the high sea and wind conditions made it evident to everyone that this means of recovery would be extremely difficult, and even dangerous.

After due consideration, it was decided to use an alternative technique, whereby a 600-foot-long nylon rope would be taken out to the floating capsule by the whaleboat. The line would be attached to a special snap-lock that was in turn attached to the head of a Shepherd's Crook. As the whaleboat came abeam of the capsule, a crew member would snap-lock the line onto the top of the craft. The crew of *Kearsarge* would then take over. The nylon line was next passed through a block near the head of the ship's crane and hand tended. As the line was hauled in it would draw the capsule alongside the carrier, placing it just below the crane hook. A special NASA nylon hoisting strap would then be put on the hook, lowered, and attached to the capsule. The capsule was then to be hoisted aboard by the crane and placed on the number 3 elevator. A pallet was constructed that would fit the contour of *Sigma 7*'s heat shield. This could also be raised by a fork-lift vehicle. Lowering the dummy capsule – impact bag deployed – onto the pallet was also practiced, as was picking up the pallet and moving it inboard.[2]

Hoisting the dummy capsule aboard. (Photo: U.S. Navy)

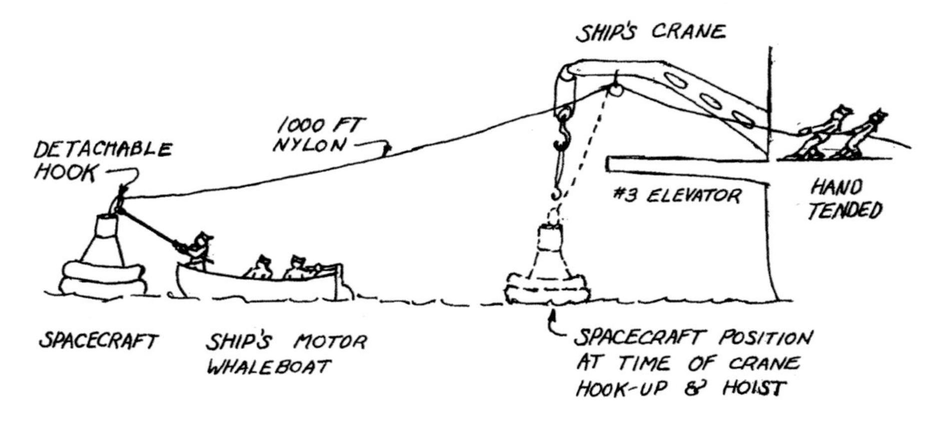

A drawing from the ship's report shows the method used in retrieving the capsule. (Drawing: U.S. Navy/USS *Kearsarge* ship's report)

In June 1962, four helicopter crews from HMM-161 (Marine Medium Helicopter Squadron 161) had been selected to train for the alternative means of recovering the spacecraft and astronaut in the event of the splashdown taking place a considerable distance from *Kearsarge*.

The prime helicopter crew consisted of:

- Capt. Kermit W. Andrus (Pilot)
- WO-1 Eugene J. Ockuly (Co-pilot)
- Sergeant José G. Abeyte (Crew Chief).

This crew would carry a three-man UDT (Underwater Demolition Team). Two teams of Navy frogmen from UDT-1 at the Amphibious Base in Coronado, California under Cdr. Wendell Webber had been in training since late July.

The prime UDT-1 team consisted of frogmen Eugene L. Dunn (commander), Gene M. Gagliardi Jr., and Billy Garner; their backups were Leonidas F. Hamel, Thomas F. Allen, and Lloyd J. Savoie.

Their training, conducted day and night (but mostly at night) involved jumping from Helicopter Utility Squadron 1 aircraft, then under the leadership of Cdr. William Casey, off Silver Strand beach. Each team would practice attaching a flotation collar around a dummy Mercury spacecraft that had been delivered there for training purposes. In order to accomplish this, the helicopter had to hover just above the water's surface. The UDT frogmen would then drop an inflatable Stullken collar (named for its designer, Donald Stullken, Chief of NASA's Recovery Operations Branch), ready to position around the spacecraft to keep it afloat, and then jump in themselves. If all went to plan, within five minutes the UDT team would have the collar secure and fully inflated around *Sigma 7*. They would then assist Schirra in leaving his spacecraft. On a signal from the frogmen the helicopter recovery could begin.

The backup crew, which would also carry a team of three UDT frogmen, consisted of:

- Capt. Phillip P. Upshulte (Pilot)
- Capt. Virgil R. Hughes (Co-pilot)
- Sgt. James W. McSavaney (Crew Chief).

One year earlier Marine Capt. Upshulte, from Quincy, Illinois, had also flown the backup Sikorsky HUS-1 helicopter involved in the recovery of astronaut Gus Grissom following the Atlantic Ocean splashdown of his *Liberty Bell 7*. During a few dramatic minutes, the pyrotechnic hatch unexpectedly blew from the floating capsule, causing it to rapidly fill with sea water and Grissom to make a hasty egress. The prime helicopter commanded by Capt. Jim Lewis had hooked onto the wallowing craft in an ultimately vain attempt to prevent it from sinking. Meanwhile, Phillip Upschulte's Sikorsky had lowered a horse collar sling to the near-drowned astronaut and hauled him aboard. On that occasion, America had come perilously close to losing an astronaut.

Also involved in the recovery training for *Sigma 7* were two other helicopter crews. As pilot of the third helicopter, Lt. Col. Lyle Tope, the skipper of HMM-161, would act as tactical air coordinator for all three aircraft, and carry a NASA anesthesiologist and official NASA photographer. This crew was made up of:

- Lt. Col. Lyle V. Tope (Pilot)
- Capt. Donald A. Dalrymple (Co-pilot)
- Sgt. John C. Thompson (Crew Chief).

A fourth helicopter crew would act in a standby capacity, ready to take over in the event of a problem affecting any of the three helicopters. This fourth HMM-161 crew consisted of:

- Capt. Allen K. Daniel (Pilot)
- 1st Lt. Gary W. Williams (Co-pilot)
- Sgt. Jimmy L. Garoutte (Crew Chief).

"Our plan is fairly simple," Lt. Col. Tope told *Windward Marine* magazine at the time. "If everything goes according to schedule, Captain Andrus will make the pickup and he and I will return to the *Kearsarge*, leaving Captain Upschulte on station until the UDT men and the capsule are picked up by a destroyer. Should the astronaut be injured, I shall lower the anesthesiologist into the water. He'll determine whether the astronaut should stay with the capsule and wait for the destroyer or whether he can be picked up. If he signals pickup, I'll come back around and hoist him aboard."[3]

Retrieval exercises off Cape Canaveral. (Photos: U.S. Navy)

Following an initial training session that involved simulated pickups in Kailua Bay, Hawaii, the four teams journeyed to Cape Canaveral in early August, where they were introduced to Wally Schirra. While there, they worked with the astronaut in perfecting rescue techniques in the Atlantic using the same techniques they would employ in the Pacific. Then it was back to their base at Kaneohe Marine Corps Air Station (MCAS) in Hawaii and rehearsing pickups in Kailua Bay until what had already become routine now became second nature.

As later described by USMC Lt. Col. Gary W. Parker, "The four crews and Captain Daniel left MCAS Kaneohe at 1600 on 28 September 1962 and landed on board the USS *Kearsarge* at 1710. The *Kearsarge* was already en route to the recovery station, with the pickup [now] scheduled for 3 October. On the evening prior to the recovery, a message was received from Commander Schirra stating that he wished to remain in the capsule until it was picked up by a destroyer, and that the final decision would be made after impact. The HMM-161 detachment was briefed accordingly."[4]

Sailors aboard *Kearsarge* form up to spell out the word "Mercury" as the carrier makes its way to the ship's designated recovery station. (Photo: U.S. Navy)

On Friday, 28 September, *Kearsarge* prepared to set sail for her assigned recovery area. Reporting aboard for the deployment was NASA's medical team of Cdr. Max J. Trummer, MC, USN (surgeon); Maj. Frank L. Mahan, MC, USAF (anesthesiologist); MSgt. R. Timon (operating room and blood bank technician); and SSgt. C. L. Stewart, USAF (EEG technician). They would later be joined by Schirra's personal physician, Capt. Richard Pollard, MC, USA. The team, designated NASA-1, brought with them 65 cubic feet of medical equipment packed in metal cases.

The carrier arrived on station on 2 October, ready to help make history.

Now in position, ready to carry out the recovery of *Sigma 7*, the carrier's crew awaits the much-anticipated splashdown. (Photo: U.S. Navy)

WITNESS TO HISTORY

For 25-year-old Lt. (j.g.) Bruce Owens on board the USS *Kearsarge*, Wednesday, 3 October 1962, was a day filled with unparalleled excitement and an overwhelming pride in his nation and its space exploits. He was a member of HS-6 (Helicopter Anti-Submarine Squadron 6) and his squadron had been placed on standby to pick up Schirra in the unlikely event that *Sigma 7* happened to splash down more than 100 miles from *Kearsarge*. Their Sikorsky helicopters had been carefully prepared for this eventuality. If, however, the splashdown was within 100 miles, a unit of Marine Corps helicopters would take off from the carrier and control the recovery.

> We left San Diego aboard the USS *Kearsarge* (CVS-33) on 1 August 1962. On board, as I recall, were two Sikorsky SH-3A Sea King helicopters and eight Sikorsky HSS-1N Seabat aircraft. The squadron's Executive Officer and Operations Officer were the senior officers in our roughly 20-officer detachment. I'm guessing we had 100 to 150 enlisted men as well. We also had a USMC detachment equipped with H-34 Sikorsky aircraft. The plan was for the Marines to do the actual astronaut recovery if the capsule landed close to the carrier, and the Navy's SH-3A would be used if its longer range capability was required.
>
> We had Navy frogman teams that would ride in the recovery helo to the splashdown location. Their job was to attach a flotation collar to the capsule, signal the astronaut when it was safe to blow the hatch, and assist him into the horse collar to be hoisted into the helo.

Bruce Owens (center) is flanked on the right by the commander of an H-34 Sikorsky, and on the left by another helicopter crewman. (Photo: Bruce Owens/U.S. Navy)

> Within our squadron the initial plan was for the two senior officers to be the SH-3A pilots, but because the shot was delayed for so many weeks it was decided that whoever was on the flight schedule that day would be the recovery crews. By luck of the draw I was the squadron duty officer the day of the recovery. This meant I was supposed to be below deck and would miss the recovery.

As Owens recalled, the carrier was also carrying reporters and photographers ready to cover the event, and the mood aboard ship that day was an excited anticipation. It began in the early hours of the morning, as he later recalled. "The night before Schirra came down, I was awakened at 1:15 a.m. by an announcement over the ship's public address system. It said Astronaut Walter Schirra had successfully left Cape Canaveral and was on his way to the *Kearsarge*. You could hear the cheers throughout the ship."[5]

Even if the Navy's Sikorskys had been called to duty, Owens – as duty officer for the unit – would have remained below decks and not participated as a co-pilot in that activity. It was a small disappointment on an otherwise sensational day. News later came through that Schirra had begun his re-entry and the landing was projected to be quite close to the carrier.

In the official (and later declassified) USS *Kearsarge* report on the recovery of the MA-8 astronaut and spacecraft, there is mention of the mounting excitement aboard the carrier following the announcement that the spacecraft had been contacted by radar and would be approaching from the starboard quarter to descend on the port bow. "Every crew-member whose duties would permit was on deck and heads shifted to monitor this indicated position in the blue sky. The press was ready with their equipment; some looked like caricatures of the American tourist abroad with three or more cameras hanging from their necks, others shouting directions to the sailors who were assisting in moving the massive telephoto equipment or lugging still more equipment and boxes of film."[6]

When it became evident that HS-6 would not be needed for the recovery, the crews were stood down from duty. Bruce Owens promptly handed over to his assistant and made his way topside to witness the landing and recovery of *Sigma 7*. It seemed like everyone was crowded into every vantage point. "All around me were reporters from the Associated Press, UPI, *Life* magazine and others," he recalled. "I took my camera and fought my way up to the second level of the ship's superstructure. Surrounded by press people, I was directly above where I thought the helo would land with the astronaut aboard. At 10:15 a.m. [Hawaiian time] it came over the PA system: '*Sigma 7* has been picked up on the ship's radar. It's 150 miles away in a descent trajectory heading directly toward our ship.'"

The first sighting of *Sigma 7* on the ship's radar screen was made by 23-year-old Ensign John A. Stephenson from San Francisco. He first spotted the spacecraft on his screen at a range beyond 150 miles and began tracking its downward flight. It was the first time ever that an American radar had picked up a re-entering spacecraft at such a distance.

Eager eyes scanned the sky in the direction that the descending capsule should be sighted. The weather was close to perfect, with light winds, a bright Sun, and a few wispy clouds. Then a collective cry went up; arms were raised, fingers pointed and excited shouts erupted all over the ship. "Next came the vapor trails overhead," said Owens. "As I looked up I saw a beautiful white arc in the sky stretching from astern of the ship to our bow."[7] A double sonic boom rumbled across the ocean.

PACIFIC SPLASHDOWN

Outlined against a high, billowy cloud, *Sigma 7* dropped from the sky on the port beam of the prime recovery ship. At 40,000 feet the small braking parachute deployed. This would steady the capsule as it descended through 15,000 feet. At around 10,500 feet, the antenna fairing above the spacecraft's cylindrical section was jettisoned, at which point Schirra said he activated the 63-foot-wide orange and white ringsail-type main parachute, which billowed out at around 10,000 feet. "Then my parachute came into view – truly a beautiful sight. If the chute hadn't appeared, of course, the day would have been a bust. I was elated, and I radioed Shepard, 'I think they're going to put me on number-three elevator.' I was of course referring to an elevator on *Kearsarge*, the recovery carrier."[8] Following the deployment of the main parachute, the impact bag also dropped from the base of the spacecraft to cushion the blow of hitting the water.

At the end of her six-orbit journey, *Sigma 7* descends by peppermint-striped parachute before splashing down right on target in the Pacific. (Photos: NASA)

Sigma 7 splashed into calm Pacific waters at 4:28:22 p.m. EDT with what Schirra would later describe as a “plop.” As expected, the spacecraft was briefly submerged before floating back up to the surface.

Recovery helicopters are quickly on the scene, hovering above *Sigma* 7. (Photo: NASA)

In his official pilot's report for NASA, Schirra said he kept a careful watch for any problems after splashing down. "On landing, *Sigma 7* seemed to sink way down in the water. It also seemed as if I were horizontal for a while. I allowed the main parachute to be jettisoned by punching in the main-parachute disconnect fuse. Then, I actuated the recovery aids switch to the manual position. The spacecraft seemed to take a long time to right itself, but again time is merely relative, and in actuality, the spacecraft righted itself in less than one minute. When *Sigma 7* had finally started to right itself, it was a very, very pleasant feeling, and at this point I knew I could stay in there forever, if necessary."[9]

Approximately 45 seconds after impact, Schirra jettisoned the main parachute so that it would not interfere with the capsule in the water. Onboard electrical equipment then shut down as scheduled, and location aids were activated. This included discharging a phosphorescent green dye marker, normally used to aid a spotting Skyraider aircraft; a Seasave beacon, a SARAH (Search and Rescue and Homing) beacon, a flashing light, and a SOFAR (Sound Fixing and Ranging) underwater charge that was dropped into the water prior to splashdown to detonate at a depth of 3,500 feet in order to give a bearing for sonar detectors aboard the task force ships.

Conditions aboard the spacecraft remained dry and relatively comfortable as *Sigma 7* finally stabilized in the ocean swells. Schirra later reported that through his window he could see the bright green marker dye spreading around the capsule. He was aware that the whip antenna pole had fully deployed, and later joked that he might have even tried to spear another bluefish.

This pinpoint landing was a truly superb feat of precise navigation which scientists and engineers had not thought they were capable of achieving at this early stage of the space program. But true to form, Schirra had proved them wrong.

Despite having joked that *Sigma 7* made "a neat aircraft and a lousy boat," Schirra opted to stay aboard his compact spacecraft, preferring to be towed to the carrier and hoisted aboard. "I had had a bad experience in ocean survival training, being lifted from a life raft to a helicopter," he later revealed. "I had bonked my head and almost fallen out of the sling."[10]

On board *Kearsarge*, Bruce Owens joined in the celebration as the latest Mercury mission came to a successful conclusion. "His heat-scorched capsule descended into the sea directly in front of our ship – a magnificent culmination of 9 hours and 13 minutes in space. He landed four and a half miles from the *Kearsarge,* becoming the first U.S. astronaut to splash down in the Pacific. A cheer went up aboard ship that drowned out everything around us."

RECOVERY

Earlier, at 9:30 that morning, local time, the three Marine HMM-161 crews had manned their respective helicopters to await the return of the *Sigma 7* spacecraft, then expected within the hour. At 10:21 a.m. the contrail of the re-entering vehicle passed overhead. The motor whaleboat was lowered over the side and then proceeded beneath number 3 elevator, where it received the end of the 600-foot nylon rope.

Once the deployment of Schirra's main parachute was observed, the helicopters took off from *Kearsarge*, ready to recover the astronaut if needed. Capt. Kermit ("Andy") Andrus established radio contact with the astronaut as soon as his aircraft, codenamed "Swiss," had lifted off from the carrier, arriving over the bobbing spacecraft just three minutes later.

Andrus called: "Astro, this is the Swiss pilot. The carrier is about three-quarters of a mile, closing."

"Okay, pilot," Schirra responded. "I think I would prefer to stay in and have a … a small boat come along side and using your collar routine to support me and have a ship pick-up. Over."

"Roger; understand," Andrus replied. "You want small ship's boat. Will give them that right away."

As Lt. Col. Parker later reported, the UDU-1 swimmers then jumped into the water next to *Sigma 7* from Andrus's primary helicopter. They quickly fitted a bright yellow Stullken collar around the blunt end of the capsule as a flotation aid until it was able to be picked up by *Kearsarge*. Meanwhile, according to Parker:

> Lieutenant Colonel Tope, commanding officer of HMM-161 and pilot of the helicopter carrying the doctor and photographer, positioned his helicopter so that the NASA photographer could photograph the collar installation. First Lieutenant Williams's helicopter hovered to the left of the capsule to allow the Navy photographer a vantage point to record the activities.
>
> The UDU-1 swimmers were unaware of Commander Schirra's final decision to remain in the capsule and because of the noise of the hovering helicopter were unable to communicate with the astronaut. They had been instructed in this case to light two red flares, which they did. Captain Upschulte retrieved one of the swimmers and explained Commander Schirra's desires and placed him back in the water. The backup swimming team was then put in the water from Captain Upschulte's helicopter to retrieve the antenna canister and the drogue chute. Later Captain Andrus picked up these swimmers and their package.[11]

The Navy frogmen were dropped 15 feet into the water beside the bobbing spacecraft, and quickly installed the Stullken inflatable collar around *Sigma 7*. (Photo: NASA)

As the frogmen await the arrival of *Kearsarge* and the motor whale boat, they let off directional flares to guide the ship. (Photo: NASA)

Wally Schirra had a particularly fond memory from this time, which is little known yarn except to a privileged few. One of those friends, Gordon Permann, a curatorial assistant at the San Diego Air & Space Museum (which Schirra referred to as "Wally World") recalls Schirra telling him this humorous little anecdote:

> He was a true Navy man, meaning he could tell a story better than most. His favorite was recounting the recovery of his spacecraft in the open Pacific. After the adrenaline rush of blasting out of the atmosphere, there was a wonderful few minutes of solitude after the parachutes lowered Wally into the sea. Laying there on his back, bobbing around like a cork, he waited for the U.S. Navy frogmen to hook him up to the rescue helicopter. Wally was sealed in his little tin can. He could hear the waves slap against the side of his ship, and could hear the swimmers as they talked about their tasks.
>
> Wally was at peace. Then, a commotion began. With a very unexpected shriek, the combat swimmer clamored up and out of the water, clinging to the top of the tiny craft. The extra weight made it swing and roll, causing its inert pilot a small amount of concern. 'What in hell is going on out there?' he yelled through the Mercury's thin skin. The swimmer, holding on to the top of the Volkswagen-sized spacecraft blurted out, 'I just saw the biggest [censored] jellyfish on the planet! It's right under us!' Wally smiled first, then laughed out loud. In his mind's eye, he could see his brightly colored parachute, fanned out below him after it sank nearly out of sight. Looming out of the depths, the spread out panels of the circular 'chute gently swam in the light current, perfectly mimicking the undulations of a 50-foot jellyfish.
>
> Even half a century later that memory brought a laugh to his voice as he recalled the surprise and alarm in the young swimmer's voice.[12]

Secured by the frogmen, the Stullken collar keeps *Sigma 7* upright until she can be reached by the carrier’s whaleboat. (Photos: NASA)

As the Marine and Navy helicopters hovered above the spacecraft, USS *Kearsarge* was cleaving a steady path to the splashdown site at a full-speed-ahead rate of 30 knots. However, the capsule had landed so near to the carrier that Capt. Rankin soon ordered the speed reduced in order that the enormous ship would not bear down too closely on the bobbing spacecraft, which was only just visible in the light swell.

USS *Kearsarge* in close proximity to *Sigma 7*. (Photos: NASA)

As *Kearsarge* approached the floating spacecraft, the 26-foot whaleboat managed on its second attempt to attach the towing line to a recovery loop fitted atop *Sigma 7*. They had been badly hampered in this task by the wind and sea spray created by the hovering helicopters. Sailors aboard the carrier then manually heaved on the nylon line, carefully hauling the spacecraft to a position beneath the carrier's boat crane where the six-inch lifting strap was attached by a frogman. Then *Sigma 7* and the attached flotation collar were winched up, dripping volumes of water. It was later said that the massive carrier actually tilted a little as everyone on board crowded over to the starboard side to watch the recovery effort.

As this was going on, in true Navy tradition Schirra spoke by radio to Capt. Rankin aboard *Kearsarge*, requesting: "Permission to come aboard, sir?" With equal gravity, the captain responded: "Permission granted."

Just 42 minutes after the spacecraft had hit the water, its heat shield was resting on the improvised pallet of wooden boxes and mattresses on the ship's number 3 elevator as a boatswain piped the nation's newest space hero aboard and announced in a loud voice, "Commander, United States Navy, boarding!"

Lt. Col. Parker explained the final duties of the HMM-161 helicopters. "Lieutenant Colonel Tope then returned to the ship to enable the doctor and photographer to be on board when Commander Schirra arrived. The other helicopters landed 20 minutes later after the successful recovery of Schirra. HMM-161 flew for a total of just 4 hours in the recovery operations. The entire operation went smoothly and exactly as briefed."[13]

At the direction of John C. Stonesifer of NASA's Recovery Operations, Flight Operations Division the flotation collar was punctured and deflated before the capsule was lowered with its hatch facing inboard.

"Reviewing familiar videos of the recovery, I remember it was always a bit difficult to settle the Mercury on the deck due to the extended heat shield secured by numerous metal straps and the large bag that attenuates the landing," he said, reflecting upon the MA-8 recovery procedure. "I do remember, once the capsule was settled I climbed to the top of the capsule and tried to communicate with Wally through the top [where a small hatch was located]. No luck with that. We received information via some other method that he was going to blow the hatch and we should stand back."[14]

Once *Sigma 7* had been secured it was time for Schirra to explosively jettison his hatch by punching a plunger with his gloved fist. He advised the crew via ship's radio that he was about to blow the hatch, and everyone was told to stand well clear. He had to hit the plunger with around five or six pounds of fist force.

Even Schirra was surprised by the instantaneous and forceful recoil from the hatch activation plunger, caused by the gases generated as it ignited. He received a painful blow to his hand, and as had John Glenn when ejecting the hatch on *Friendship 7*, he suffered a minor injury to his hand, even though he was still wearing the glove. Later, at his medical debriefing, he was not slow to point out the tell-tale impact bruising and cut on the front of his hand. It may have caused him some minimal pain at the time, but he would later state emphatically that it vindicated his friend and fellow astronaut Gus Grissom, who was under suspicion for accidentally blowing the hatch of *Liberty Bell 7*, causing it to be lost at sea. Both Glenn and Schirra proved that it took a considerable punching blow to operate the plunger, and Grissom did not suffer any bruising or cuts on his hands.

The whaleboat crew attaches the nylon tow rope to *Sigma 7*, which is then slowly hauled towards the carrier as the boat stands by. (Photos: NASA)

Sigma 7 is hoisted out of the water by boat crane, ready to be placed on a special pallet on the carrier's number 3 elevator. Note the ship's company crowding to one side of USS *Kearsarge*. (Photos: NASA)

The Stullken collar, its job done, is punctured before the spacecraft can be lowered onto the pallet. (Photo: NASA)

With the hatch safely blown, John Stonesifer was one of those who quickly moved up the now-open capsule. "After he exploded the hatch a few of us rushed forward to check his condition then assist his egress. My most lasting memory of that moment was when I glanced in I was startled, for I thought he was injured. His face was white as a sheet. Then I saw that big grin and knew he was OK. I suppose his face was white from the residue from exploding the hatch."[15]

Sigma 7 is carefully guided onto the waiting pallet. (Photo: NASA)

This photograph captures the moment Schirra blew the hatch on *Sigma 7*. (Photo: NASA)

Once the hatch had been blown, the doctors did a quick check of Schirra to ensure that he was okay. (Photo: NASA)

Schirra is assisted out through the hatch. (Photos: NASA)

Schirra began to back head-first out through the open hatch with the aid of some willing helping hands. Then, sitting on the side of the spacecraft, he removed his white space helmet and, to the cheers of those on board, looked over the exterior of *Sigma 7,* his hair matted with perspiration. Next he stood on the sill of the hatch and reached up to touch the cylindrical upper area of the spacecraft to find out how well it had fared during re-entry. He later explained that he was inspecting the test shingles of ablative material attached to it.

The astronaut climbs to the top of his spacecraft to check on the condition of the ablative shingles. (Photos: NASA)

The shingles were panels of materials which burned away under the extreme heat of atmospheric re-entry. "Unfortunately," he later reported, "they all looked the same – they all had the same charred appearance and this would take careful analysis that the naked eye could not determine. They all looked good, I might add."[16]

With his space journey of 160,000 miles at an end, Wally Schirra stepped onto the deck of USS *Kearsarge* at 11:13 a.m.

POST-FLIGHT DUTIES

Having carried out his brief inspection, Schirra turned around with a broad grin on his face to acknowledge the cheers and applause of the ship's crew, who were peering at him and his spacecraft from every possible vantage point on the carrier. He then shook hands with the ship's captain and senior crewmembers, including Capt. Thomas King, commander of the recovery task group, and some NASA officials.

"From that point on, my team and I were quite busy and I had little time to observe Wally's waving to the ship's crew and greetings by the *Kearsarge* CO and staff," John Stonesifer recalled. "Following the greetings, he was escorted by the flight surgeons and other NASA personnel to the ship's sick bay for his post flight-examination."[17]

His hair matted with perspiration, Schirra is welcomed aboard by military and NASA officials. (Photo: NASA)

As he walked across to the elevator that would carry him up to the flight deck of the ship, reporters shouted to Schirra, asking how he felt. "Fine," he replied, with a casual wave of his hand. When his entourage arrived on the flight deck everyone continued to cheer loudly and crowded in to catch a closer glimpse of the astronaut.

After everyone had cleared the area around *Sigma 7*, the spacecraft was minutely examined and photographed by NASA technicians. It was then hoisted up once again to the full extent of the impact bag so that the entire craft could be washed down with fresh water. New padding was then provided to replace the wet ones in the pallet and *Sigma 7* was lowered down onto it again. Then the pallet, with the spacecraft secured, was moved inboard to hangar bay 3, where the sailors were permitted a close-up view. Two Marine sentries were posted to prevent anyone from touching the craft or taking photos of its interior.

A close-up view of the ablative shingles (now numbered with chalk) on *Sigma 7*. (Photo: NASA)

Schirra appeared somewhat fatigued but assured William ("Red") Hayes, the senior NASA representative on board, "I feel fine, just fine. A great trip. What a sweet little bird." The cute metaphor would become a much-quoted remark in the following day's newspapers. He then went quickly below deck for his medical examination.

Capt. King, who directed the Navy's part of the recovery, was full of praise for the recovery process and later said that "fantastically accurate navigation both by Schirra and the naval vessels involved made possible a very excellent job of rescue at sea of an important piece of cargo. The tremendous amount of information fed to us at the right time, the good weather which made possible both visual and electronic observation and the fact that the flight by Schirra was perfect; all these share credit in an unprecedented feat."[18]

Wally Schirra arrived in the ship's medical quarters at 11:20 a.m. One of the first things he did before undergoing his medical examination was proceed to the office of Capt. Nell (of the ship's medical team) to receive a long-distance call from the White House. President Kennedy had watched the launch and the ensuing commentary for about 90 minutes on television, and he was anxious personally to congratulate Schirra on his successful mission.

JFK: Hi, Commander.
Schirra: Yes sir, Mr. President.
JFK: We are delighted with your trip. I will tell you that.
Schirra: I thought I might as well go where I was headed this time.
JFK: You did a wonderful job and we are very, very pleased.

Schirra:	I appreciate your coming down and giving our booster a blessing. It helped.
JFK:	Well, it does us a lot of good, so I certainly extend all the congratulations to you and your family.
Schirra:	Thank you very much, sir.
JFK:	I will look forward to seeing you sometime soon.
Schirra:	I will look forward to it too.
JFK:	Thanks, Commander. Good luck.
Schirra:	Thank you for calling, sir.
JFK:	Right. Goodbye.

Schirra receives a congratulatory call from President Kennedy. (Photo: NASA)

Next, Schirra received a call from Vice President Johnson and then one from his joyful wife Jo and their two children, who had followed his flight by both radio and television. He assured them that everything had gone extremely well, and he was in good health and spirits. Upon completion of these formalities he was escorted to the treatment room where he was helped out of his silvery space suit. He then donned a comfortable dressing gown.

NASA photographer Gene Edmonds took photographs of Schirra in the treatment room. He had earlier been aboard one of the helicopters taking shots of *Sigma 7* being lowered onto *Kearsage*'s deck. He recalls there was one unpleasant situation involved in removing Schirra's space suit, which fortunately they all laughed off. "Wally's urine bag had burst upon the capsule's impact with the ocean, and we had lots of laughs while undressing him for the medical examination."[19]

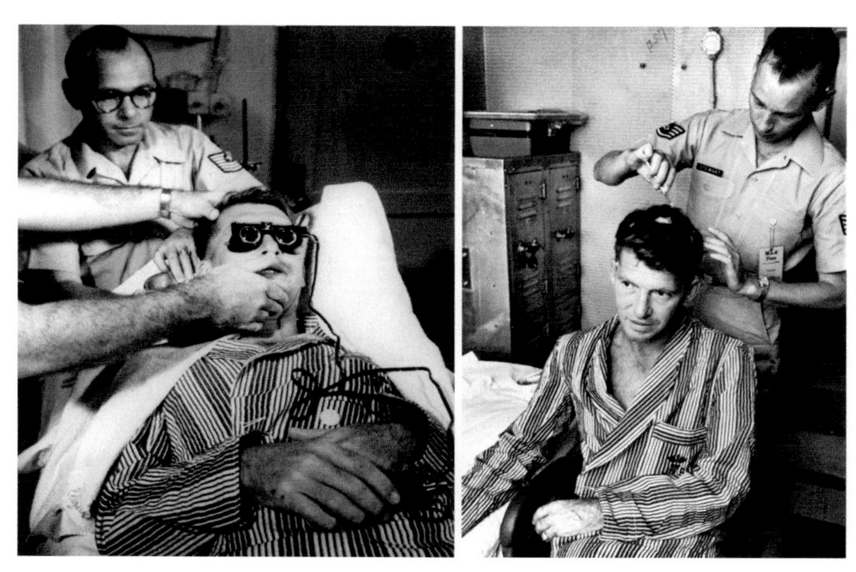

The astronaut is given a thorough medical examination after shedding his space suit. (Photos: NASA)

Schirra then had a general physical examination – vital signs, chest, ears, nose and throat. He was given an EEG and chest x-ray; he had blood taken, gave a urine sample and completed a caloric test. At 2:30 p.m. he left the sick bay and made his way to the admiral's in-port cabin for further verbal debriefing and medical observation.

More extensive tests and a thorough debriefing would take place over the next three days during a leisurely three-day trip back to Pearl Harbor. Actually, Capt. Rankin had plotted a roundabout route to Hawaii via Midway in order to give the doctors, scientists and NASA officials time to fully explore Schirra's medical condition and document his impressions of his space flight in a leisurely manner.

According to a space agency spokesman, the first medical examinations indicated Schirra to be in "excellent condition" with no ill-effects from his flight. Commander Max Trummer, a Navy surgeon on the medical debriefing team, reported the astronaut "looks as healthy as a bear."[20]

Schirra's only grouse was that during his flight he had not received the score in the San Francisco Giants-Los Angeles Dodgers final playoff game that he had requested. He would have been amused to hear that an unidentified baseball fan had earlier called the *Worcester Evening Gazette* (Massachusetts) with a request that Schirra's flight be shortened. "I want him down before the playoff game starts," the caller stated. He explained that the final playoff between the Giants and Dodgers was being televised, and "we might not be able to see the start of the game." But NBC replied it would be televising the game as scheduled.[21] For the record, the San Francisco Giants strolled into the World Series with a 6 to 4 victory over the Dodgers.

Schirra's personal physician, Dr. Richard Pollard, who worked with Dr. Trummer on the preliminary examination, said Schirra had shed four pounds in weight in making the flight, but that "was less than he would have lost playing a game of football." He also reported that the astronaut showed no immediate evidence of cosmic rays, abnormal bodily functions, or changes in his equilibrium. According to Dr. Pollard, Schirra had eaten very little during his six orbits: two containers of fruit and some liquids, but no chunks of solid food. "When you're having that much fun you just don't get hungry," he quoted Schirra as saying.[22]

That evening Schirra enjoyed a hearty dinner in his cabin of two minute steaks, grapefruit salad, mixed vegetables, iced tea, and a strawberry sundae. Apparently he was still too excited to retire and wanted to talk while maintaining his renowned sense of humor. He finally went to bed after a long and busy day, 21 hours and 40 minutes after waking. After 10 hours he woke up, used the head, talked and smoked for about an hour, then returned to bed and slept for another three hours.

The day following the recovery, the NASA-2 medical team arrived aboard the USS *Kearsarge* by airplane. This team comprised of Dr. Charles A. Berry; Dr. Howard A. Minners; Capt. Ashton Graybiel, USN; Lt. Col. D. Flynn, USAF; Dr. George Ruff; Lt. Col. Edward C. Knoblock, USA; Lt. Col. James Culver, USAF; Capt. W. B. Clark, USA, and one laboratory technician. They brought with them an additional 25 cubic feet of medical equipment.[23] Schirra returned to the sick bay for a further round of tests that included an electrocardiogram, visual acuity determination, phorias check, psychiatric evaluation, slit-lamp examination, opthalmoscopic examination, tangent screen and near-vision tests.

Capt. Kermit ("Andy") Andrus, who served as the prime recovery helicopter pilot aboard *Kearsarge* for *Sigma 7* would enjoy a long and distinguished career in the U.S. Marines, retiring after 25 years' service with the rank of colonel. Although he did not transport either the spacecraft or the astronaut onto the waiting carrier, he still recalls it as "a great day." Especially when it came time to celebrate the historic occasion. "A group of people who were deeply involved in the mission was bought aboard the ship shortly after the spacecraft landed – this was a very happy time. I smile remembering this. It is known that liquor is not to be brought aboard ship – however, it appeared – quite an amount of it! The group, of which I was a member, decided that rules or not, we would have a congratulatory toast (or a few toasts). We then took on the name of the 'Bad Ass Club' (for bringing the liquor on board). I was proud to be a member. Wally presented me with a picture of himself signed: 'Andy, thanks for being my Guardian Angel.' The day ended with much camaraderie."[24]

SETTING SAIL FOR HAWAII

Soon after the recovery of Schirra and his spacecraft from the sea, *Kearsarge* had set sail southwest toward Midway Island some 275 miles away, where *Sigma 7* would be transferred to a tug at first light for transportation to the airport, ready for being flown back to Cape Canaveral on a Lockheed C-130 Hercules cargo plane for examination by engineers and technicians.

Schirra was to spend the sailing time in the comfort of the admiral's quarters. On first entering his quarters, he roared with laughter at a little memento waiting for him from Dee O'Hara. "Hanging in the head was an oversized urine collection device, a duplicate of the one she had presented to me before the flight. Dee had given it to one of the doctors to bring all the way from Cape Canaveral."[25]

Dee recollects, "As to smuggling that urine collection 'bag' on board, I gave it to one of the recovery people to take aboard and see that it got displayed above the head before Wally was piped on board. I was surprised and pleased that the mission was accomplished. I heard that Wally loved it and they could hear him laughing all over the ship."[26]

The morning after his recovery, Schirra was up at 7:00 a.m., cheerful and in fine spirits. After an eight-hour sleep he enjoyed a breakfast of scrambled eggs, sausages and four cups of coffee. One of his first comments of the day was typically Schirra, "Hose down *Sigma 7*; I'm ready to go again!"[27]

With breakfast completed, Schirra was surrounded by engineers, technicians and tape recorders as he sat at a mahogany table in the admiral's quarters and recounted his epic voyage through space. He recorded on tape his recollections of the previous day, giving every detail from the preparations for launch at Cape Canaveral through to the bull's-eye landing near USS *Kearsarge* some 9 hours and 13 minutes later. After that he was free to move around, but was under strict instructions from NASA that he was not to make any statements to the press while on board the carrier. He would give his story to the press and public at a Sunday news conference back in Houston.

In the Soviet Union, the latest Mercury flight was relegated to the back pages of the Russian press, which instead focused on the fifth anniversary of the historic launch of the first *Sputnik* satellite, along with broad hints that there might soon be another Soviet space spectacular. In fact this would not occur for several months, when Vostok-5 and Vostok-6 were launched two days apart in June 1963 with the latter spacecraft carrying Valentina Tereshkova as the first woman to fly into space. A Moscow Radio home service broadcast did describe Schirra as "a courageous son of the American people," but stressed that Soviet cosmonauts had thrice flown in space for longer durations.

The official Soviet news agency, TASS, indicated that one or more of its four space veterans might soon take off on a second orbital flight. "It is quite likely," said TASS, "that in the comparatively near future some of the present quartet of cosmic brothers will once again join a subgroup of immediate flight readiness. A man travelling in outer space for the second time will be able to see and understand there much more than a newcomer."[28] In fact the first re-flight of a cosmonaut did not occur until 1967 and it ended tragically when Col. Vladimir Komarov was killed striking the ground at high speed after the parachute system failed on the troubled inaugural mission of the new Soyuz spacecraft.

Once the greetings and congratulations were over, Schirra sat down with Walt Williams (left) and his fellow astronauts to talk about his flight. (Photo: NASA)

At Midway, five Mercury astronauts and Project Mercury director Walt Williams boarded the NASA-2 medical team's COD (Carrier Onboard Delivery) aircraft for a flight over to *Kearsarge* and a laughter-filled reunion with Schirra. This gave them a chance to question him about different aspects of his flight while the ship headed for Hawaii. The only Mercury astronaut not able to be there was Alan Shepard, who was still on duty aboard the Pacific Command Ship *Rose Knot*, stationed off Guam. Once *Kearsarge* was several miles from Pearl Harbor on Saturday, Schirra was scheduled to be flown to Hickam Air Force Base for a brief 10-minute stopover prior to boarding a flight to Houston. That, at least, was the plan at the time – but things would change.

Schirra was particularly eager to share some news with Gus Grissom. In *Schirra's Space* he wrote, "I blew the hatch on purpose, and the recoil of the plunger injured my hand – it actually caused a cut through a glove that was reinforced by metal. Gus was one of those who flew out to the ship and I showed him my hand. 'How did you cut it?' he asked. 'I blew the hatch,' I replied. Gus smiled, vindicated. It proved he hadn't blown the hatch with a hand, foot, knee or whatever, for he hadn't suffered even a minor bruise."[29]

For Bruce Owens, even though the ship's crew were not permitted to talk to Wally Schirra, it nevertheless presented a unique opportunity to meet some of his heroes. "I talked with John Glenn in the officers' wardroom for quite some time one night on the trip back. It was an exciting time for a 25-year-old junior grade lieutenant."[30] He would eventually retire with the rank of captain in 1982.

Bruce Owens with his wife Nonnie in 2014. (Photo: Bruce Owens)

In referring to the space flight, a delighted Project Director Walt Williams told the reporters that "as far as I'm concerned it was perfect." He said the flight prepared the way for an 18-orbit, 24-hour flight the following year. Williams added that three Atlas boosters and three capsules would possibly be allotted to further Mercury missions, but he did not know whether they would all be used before an attempt was made to orbit a [Gemini] spacecraft carrying two astronauts.

Famous for his own "gotchas," Schirra later revealed he was on the receiving end of yet another one during his six-orbit flight, this time hatched by fellow astronaut Gordon Cooper and good friend Jim Rathmann, an Indy car racer. As Schirra's backup Cooper had easy access to *Sigma 7* prior to launch, and he had concealed behind the instrument panel a little in-flight surprise for the pilot. A trimmed-down Tareyton packet held four cigarettes, but that was not the only bit of contraband to go into the little hiding place, because there was also a miniature bottle of scotch whiskey. Wally would later say of this prank, "Before I flew on Sigma 7, Gordon Cooper and Jim Rathmann … arranged to put a miniature, airplane catering-sized bottle of Cutty Sark scotch and one row of Tareyton cigarettes way in the back of a compartment in the instrument panel … I drank the Scotch as soon as I had a chance on board the recovery vessel. The medics all wondered why I had a small alcohol level in my post-flight blood tests! I guess a nicotine level would've really thrown them!" He kept the cigarettes as a reminder of that humorous "gotcha" played on him.[31]

News representatives are able to discuss Schirra's flight with Walt Williams and five other Mercury astronauts in front of *Sigma 7*. (Photo: NASA)

HONOLULU TO HOUSTON

After three days of debriefing aboard *Kearsarge*, Schirra and his fellows departed for home at 9:45 a.m. on 6 October, leaving the carrier aboard a Grumman Tracker COD aircraft (with Schirra acting as co-pilot). However their planned transit in Hawaii was extended by three hours to permit a reception and lunch at Hickam AFB. Apparently there had been some bruised civic pride in Honolulu and pressure had been applied to NASA by Hawaiian Governor William F. Quinn and Mayor Neal Blaisdell to change the schedule and make a formal welcome in Honolulu possible. Eventually, the space agency relented and hurried plans were put in place.

On Saturday, 6 October at 10:02 a.m. (Hawaiian time), Wally Schirra set foot on American soil once again at Hickam. A crowd of about 2,000 was there to greet him. Half were military personnel and the rest were civilians dressed in colorful Hawaiian shirts and muu-muus. They cheered and applauded as America's newest space hero stepped up onto a flatbed truck for some welcoming ceremonies. He was followed on the platform by his five fellow astronauts.

Schirra receives a ceremonial lei from Kalani Flood after his arrival at Honolulu airport. (Photo: NASA)

In accordance with Hawaiian tradition, Schirra was given a double lei of bright red carnations and received a kiss from eight-year-old Kalani Flood. A tag on the lei read: "From the thousands of little people whose dimes and nickels built *Sigma 7*, and who very proudly bask in your reflected glory." The group of astronauts then waved to the excited crowd.

The airfield ceremonies only lasted 15 minutes before the astronauts and the official party were ushered into military limousines for the drive through light rain to a private luncheon hosted by Pacific Air Forces commandant Gen. Emmett O'Donnell, Jr. At the function Schirra also received the congratulations of Governor Quinn.

An old friend happened to be in Hawaii – the comedian Bill Dana, whose hilarious routines as a reluctant astronaut had caused him to become a close friend of the actual astronauts; so much so that he was known to them as the Eighth Astronaut. Learning that Schirra was in town for a brief stopover, Dana put in a phone call to the astronaut and after a lot of difficulty was finally put through. As Schirra tells it, Dana said, "Hi Wally, how are you?" To which he replied, "Oh, I've been around." Then he hung up, bursting into laughter. He later pointed out that giving Dana one of his better puns was simply too good an opportunity to miss.

Schirra and Walt Williams enjoy some Hawaiian hospitality over lunch. (Photo: NASA)

Sigma 7 is offloaded from the carrier for transportation to Cape Canaveral. (Photo: NASA)

Sigma 7 returns in glory to Hangar S at the Cape. (Photo: NASA)

Once the lunch and formalities were at an end, the astronauts boarded a Military Air Transport Services jet at 1:19 p.m. for a non-stop flight to Houston, arriving in the early morning hours of Sunday.

Reminiscing on that excruciatingly long flight for the author in 2003, NASA public affairs officer Paul Haney said, "I recall in that 15-hour flight back to Houston in a slow KC-135, John [Glenn] was writing in long-hand every time I woke up. At one point I asked him what he was doing [but] he wouldn't let me read it. John was a documenter. But Wally was quite at ease with himself and quite chatty. His Mercury flight was a 'textbook flight' according to the engineers in charge. Wally came down about three feet from the planned spot.

Unlike John, there were no false signals or warnings of landing bag deployments. Everything was nominal. We kidded Wally that if he had just strayed a little west in his landing, then he could have achieved the first 24-hour Mercury flight by landing on the other side of the dateline."[32]

Meanwhile the USS *Kearsarge*, her job accomplished, re-entered Pearl Harbor at 11:00 a.m. on 6 October. On her starboard side a huge banner was proudly displayed with letters six feet high proclaiming: "CANAVERAL TO KEARSARGE."[33]

REFERENCES

1. USS *Kearsarge* (CVS-33), (declassified paper) "Project Mercury Report of MA-8, 1 August – 17 October 1962"
2. *Ibid*
3. GySgt. Jack T. Paxton, *Windward Marine* magazine article, "Pick-up Teams for Project Mercury," issue Vol. II, No. 39, 28 September 1962
4. Lt. Col. Gary W. Parker, USMC, *A History of Marine Medium Helicopter Squadron 161*, History and Museums Division, Headquarters, U.S. Marine Corps, Washington, D.C., 1978
5. Bruce and Nonnie Owens email correspondence with Colin Burgess, 23 May 2015 – 9 November 2015. Incorporates Owens quotes from Don Moore "War Tales from the Front," online at: *http://donmooreswartales.com/author/talesfromthefront*
6. USS *Kearsarge* (CVS-33), (declassified paper) "Project Mercury Report of MA-8, 1 August – 17 October 1962"
7. Bruce and Nonnie Owens email correspondence with Colin Burgess, 23 May 2015 – 9 November 2015. Incorporates Owens quotes from Don Moore "War Tales from the Front," online at: *http://donmooreswartales.com/author/talesfromthefront*
8. Walter M. Schirra, Jr. with Richard N. Billings, *Schirra's Space*, Quinlan Press, Boston, MA, 1988
9. Walter M. Schirra, Jr., *Pilot's Flight Report: Results of the Third United States Orbital Space Flight, October 3, 1962*, (ASA SP-12) Office of Scientific and Technical Information, NASA, Washington, D.C., December 1962
10. Walter M. Schirra, Jr. with Richard N. Billings, *Schirra's Space*, Quinlan Press, Boston, MA, 1988
11. Lt. Col. Gary W. Parker, USMC, *A History of Marine Medium Helicopter Squadron 161*, History and Museums Division, Headquarters, U.S. Marine Corps, Washington, D.C., 1978
12. Gordon Permann (San Diego Air & Space Museum) email correspondence with Colin Burgess, 25 June 2015
13. Lt. Col. Gary W. Parker, USMC, *A History of Marine Medium Helicopter Squadron 161*, History and Museums Division, Headquarters, U.S. Marine Corps, Washington, D.C., 1978
14. John C. Stonesifer email correspondence with Colin Burgess, 24 June – 2 July 2015
15. *Ibid*
16. NASA *Space News Roundup*, "Schirra Talks On Mission," MSC, Houston, issue 17 October 1962, pp. 1–2

17. John C. Stonesifer email correspondence with Colin Burgess, 24 June – 2 July 2015
18. *Chicago Daily Tribune* newspaper, unaccredited article, "Schirra Fished Out of Ocean; Trip 'Perfect'," issue Thursday, 4 October 1962, pg. 12
19. *The Memoirs of Eugene G. Edmonds*, NASA Johnson Space Center Oral History Project, Appendix B. Houston, TX, 11 December 2003
20. *Chicago Daily Tribune* newspaper, unaccredited article, "Schirra Fished Out of Ocean; Trip 'Perfect'," issue Thursday, 4 October 1962, pg. 12
21. *The Daily News*, St. John's (Newfoundland), "Baseball Not Orbits," issue 4 October 1962, pg. 6
22. *Evening Independent* (Peterborough, FL) newspaper article, "Schirra 'Happy as Lark,'" issue 4 October 1962, pg. 2A
23. USS *Kearsarge* (CVS-33), (declassified paper) "Project Mercury Report of MA-8, 1 August – 17 October 1962"
24. Col. Kermit Andrus email correspondence with Colin Burgess, 7 July 2015
25. Walter M. Schirra, Jr. with Richard N. Billings, *Schirra's Space*, Quinlan Press, Boston, MA, 1988
26. Dee O'Hara email correspondence with Colin Burgess, 26 June 2015
27. The *Spokesman-Review* newspaper (Spokane, Wash.) unaccredited article, "Schirra Ready To Go Again," issue Friday, 5 October 1962, pg. 1
28. *St. Petersburg Times* (Florida) newspaper, unaccredited article, "Schirra Begins Leisurely Journey Back to the U.S.," issue Friday, 8 October 1962, pg. 10-A
29. Walter M. Schirra, Jr. with Richard N. Billings, *Schirra's Space*, Quinlan Press, Boston, MA, 1988
30. Bruce and Nonnie Owens email correspondence with Colin Burgess, 23 May 2015 – 9 November 2015. Incorporates Owens quotes from Don Moore "War Tales from the Front," online at: *http://donmooreswartales.com/author/talesfromthefront*
31. RR Auctions promo material, *Wally Schirra's 'Gotcha' cigarettes, discovered during his Sigma 7 flight*, January 2013. Online at: *http://www.rrauction.com/101_wally_schirra_gotcha_cigarettes.cfm*
32. Paul Haney email correspondence with Colin Burgess, 30 September 2003
33. *St. Petersburg Times* (Florida) newspaper, unaccredited article, "Schirra Begins Leisurely Journey Back to the U.S.," issue Friday, 8 October 1962, pg. 10-A

7

Press conferences, parades, and post-flight honors

Following the seemingly endless flight across from Hawaii, Schirra's transport aircraft finally touched down at Houston International Airport shortly after one o'clock in the morning. Once the engines had shut down the forward door was opened and a tired but nevertheless exhilarated astronaut descended the steps to be greeted by Governor Price Daniel, the Mayor and Mrs. Lewis Cutrer, assorted NASA officials, and a gathering of beaming politicians from Washington and Texas. His family was there, eager to see him, and they were surrounded by other astronauts and their wives, with a small but happy crowd of well-wishers standing farther back. It was a heartwarming and very welcome homecoming for Schirra.

Once he had hugged and kissed his family, presented Jo with his Hawaiian lei, and shaken hands with all the dignitaries and his friends, Schirra received a quick briefing from a NASA official. He was informed that later in the day he would be honored by a parade through the city that would conclude with ceremonies at the memorial center on the Rice University campus.

Despite the early morning hour, a motorcade had been arranged to take Schirra back to his newly completed home in Timber Cove, south of Houston. Due to the intensity of his training program he had been unable to devote much time to his family, and had never set foot inside the house. But after everyone was inside, he had time for a quick look around before the champagne corks began to pop. It was finally time to relax and celebrate with his family and friends.

The partying continued to 4:00 a.m., at which time everyone decided it was time to let their hero get some rest. They straggled out, leaving the family in their new home. Wally Schirra may have been the world's newest spaceflight hero, but he was quickly brought down to Earth in a hurry after they'd cleaned up when Jo asked him to put out the garbage.

"What I had done to Bill Dana in Hawaii was nothing to what my wife had just done to me," he recalled. "The ultimate put-down."

C. Burgess, *Sigma 7*, Springer Praxis Books, DOI 10.1007/978-3-319-27983-1_7

On that day, the people of Houston began turning out in their thousands to give the astronaut and his family a huge, friendly Texas homecoming. Early that afternoon the astronauts and their families gathered at the Houston headquarters of NASA's Manned Spacecraft Center, then temporarily based in the Farnsworth & Chambers building on South Wayside. The children of the astronauts were delighted to meet Jay North, the young star of the hit television show *Dennis the Menace*, who was in Houston to kick off a savings bond drive. After lunch the motorcade began, proceeding slowly down Telegraph Road to South Wayside. From there it would go over to the Gulf Freeway and on to Rice University.

Later police estimates placed the crowd in downtown Houston at around 300,000 citizens who – despite the hot weather and high humidity – cheered and applauded as the motorcade wound slowly through the city.

With his family seated in the rear of the leading limousine, Schirra was perched atop the back of the seat, waving to the crowd. Behind them came the other astronauts, with the exception of Alan Shepard, who was still to return from his post aboard the Pacific tracking ship, *Rose Knot*. Farther back in the parade were NASA Administrator James Webb, Congressmen Albert Thomas, Olin Teague and Bob Casey, Dr. Robert Gilruth, director of the Manned Spacecraft Center, and various other MSC officials. Eventually the motorcade rolled into the campus of Rice University.

NASA's temporary Houston office facilities at the Farnsworth & Chalmers building, from where the motorcade began. (Photo: NASA)

Young television star Jay North kitted out in astronaut garb with the Schirras before the motorcade moved off. (Photo: NASA)

The sign says it all for Wally Schirra. (Photo: NASA)

Escorted by motorcycle police, the motorcade continues through the streets of Houston. (Photo: NASA)

William Marsh Rice University, more commonly known simply as Rice University, is a private research university founded in 1912 with its campus in Houston, Texas. On that afternoon of 7 October 1962, it played host to the world's newest spacefarer at a nationally televised post-flight press conference.

As NASA's Public Affairs Officer Paul Haney explained, "We held the press conference at Rice to acknowledge the role of the Rice Board of Trustees in making land available for the new Manned Spacecraft Center. Rice also knew that in deeding the land to NASA they could potentially realize some considerable government grants and education recognition down the line."[1]

THE "TEXTBOOK" FLIGHT OF *SIGMA 7*

The press conference kicked off at 4:00 p.m. in the auditorium of the Rice University Memorial Center, with Schirra seated and flanked on stage by James Webb and Robert Gilruth. NASA Press Officer "Shorty" Powers stood to one side, ready to step in and answer any questions about the space agency and its plans which might prove difficult for Schirra.

The motorcade sweeps by the iconic Metropolitan Theater on Main Street, which was demolished 11 years later. (Photo: NASA)

Webb introduced himself, Gilruth, and Schirra to the 300 guests and then gave some opening comments and news. He announced that Schirra and his family would travel to Oradell, New Jersey for a homecoming parade that would be held on 15 October. Also, on 25 October, NASA's annual awards ceremony would be held in Washington, D.C.

In talking about Wally Schirra and his recent mission, Gilruth pointed out that flights like MA-8 were essential if America was to progress in space, and required the greatest of efforts on the part of many individuals and organizations. He also warned that these were pioneering flights that involved considerable elements of danger.

Robert Gilruth, Wally Schirra and James Webb at the Rice University press conference. (Photo: NASA)

The press conference was a massive media event. (Photo: NASA)

Before he began to discuss his mission, Schirra took a few moments to introduce his parents, wife and son, who were seated near the front. Their understandably exhausted daughter Suzanne had already gone home with Jo's mother. Schirra also acknowledged the presence of five of his fellow Mercury astronauts, with Alan Shepard still absent on duty.

In anticipation of the question being asked by reporters, Schirra then explained the reasoning behind his choice of the name *Sigma* for his spacecraft. "*Sigma 7* is a name to me that connotes an engineering symbol. It is the eighteenth letter of the Greek alphabet, and it connotes summation... Basically, what we wanted to connote with this name was the very many inputs that have been brought forth to develop this flight – the fact that we had previous flights where we needed to make our initial steps into space, the fact that operations analyses of the previous flights, engineering analyses, showed we had to make minor changes to make previous problems straighten out."

Schirra then said that the earlier Mercury flights – suborbital shots by Shepard and Grissom, and the three-orbit missions of Glenn and Carpenter – meant much to him in helping him realize what he needed to do to add to the knowledge already accumulated. He also affirmed that the sensations of launch were perfectly described by both Glenn and Carpenter. "The railroad train that you are sitting on really does move out."

In describing his activities during the early phases of his flight, Schirra referred to MA-8 as "a textbook flight." It was an appropriate term that would later be applied, and often, to his mission. He also mentioned a number of changes incorporated in the spacecraft for his mission. Attitude control, he stated, was perfect during the flight.

On another subject, Schirra emphasized the sense of cooperation between himself and the tracking stations, including the Mercury Control Center. "It was a nice feeling to realize that questions I had could be answered; questions they had, I could answer."

The small amount of maneuvering propellant he used during his six orbits was also discussed. "My intention was to use so little fuel that no one could deny that we had enough fuel aboard *Sigma 7* for eighteen orbits if we wanted it. I think I proved that point. As far as problems go, there is [one] which we have solved – this was the suit temperature. I have been much hotter in the tent at Cape Canaveral than I ever thought of being in *Sigma 7*." At one point he said he had such control over the coolant setting that he actually got cold; "the first [time] as far as I know we ever got cold in space."[2]

During his presentation, Schirra said he came back from his mission with just one suggestion for the next flight in the Mercury series – "just move up its launching date." He added that he came back from his flight reluctantly. "I would [have liked] to have gone for twelve more orbits."

Schirra revealed that at one point he totally cut himself off from ground control. In order to conserve the power in his *Sigma 7* spacecraft he threw a switch, shutting down the electrical system which would have permitted ground control stations to bring him out of orbit in an emergency. "From the beginning," he said, "I wanted to turn off the armed squib [control box], which means that it is my capsule and no one can bring me back until I put the switch back." He left it off until Flight Director Chris Kraft called and said, "I think you have proved your point, old buddy."

Asked if he had any moments of apprehension, Schirra didn't hesitate, saying, "Not one. I had no problems... no uneasiness... no queasiness. I had no fatigue... I was ready to continue through one day."[3]

The countdown to lift-off, he said, "was a real dream" in its smoothness. Asked whether he had seen the Echo balloon satellite while in orbit, which had been a pre-flight possibility, he explained that he did not see the huge, 90-foot diameter satellite, but also had not been willing to use extra attitude control to look for it. In response to another question about visibility, Schirra said the exhaust from the jettisoned escape tower mounted on the top of his spacecraft did cloud his window, but only to a slight degree. He said it was "much like you see on a dirty windshield before you get a tank of gas."

Several other points came out of the press conference. One was that a spacecraft in free flight has a very slow drift rate, rather than a "tumble." During the drifting period of his flight no single axis drift rate ever exceeded about three-quarters of a degree per second. At this rate, he explained, it would take the spacecraft six minutes to perform one revolution.

The ground-based flare set off at Woomera was not visible to Schirra through the existing cloud cover, but he did see lightning over Woomera at one time when it was not visible from the ground.

He emphasized to the newsmen that he experienced very little fatigue during his flight, and did not feel tired until some five hours after his recovery from the Pacific Ocean. He thought the necessity of sleep on long flights would be no problem, since uncontrolled drifting, used extensively on his flight, was not a problem. He also said that his only problem food-wise, was that it was stowed beyond easy reach. He hadn't experienced the same problem with food crumbling that Scott Carpenter had reported during his flight. He ate two tubes of food, specifically peaches, beef and vegetables, but in the tight confines of the spacecraft he had been unable to extract two containers of dessert-style food cubes from their awkwardly placed stowage.

The photographs which he took with his special hand-held Hasselblad camera were minimal, he explained, because he was busy during the first orbit with his suit circuit and an evaluation of control systems, and later had few opportunities to take pictures.

A dosimeter, which he explained was a radiation measuring device, had shown less than one-tenth of one roentgen of radiation present. "We did not anticipate radiation as a problem, and as it turned out, we had less than I have in my wrist watch."

In discussing the spacecraft's periscope, Schirra said, "I almost launched without a periscope on this flight and would launch again without a periscope. I did not use [it] very often. We felt that it was best to take [it] along and put to bed, once and for all, whether we needed it to acquire attitudes or whether we could [do that] exclusively with the window. I will state now that there is no requirement for the periscope."

Only one loose object was spotted drifting about the cabin during weightlessness. It was, he said, "a very minute little washer." He then added with a laugh, that "Gordon Cooper has it in custody and I am going to present it to my capsule engineer when he gets back to the Cape."[4]

Former flight controller Dutch von Ehrenfried has said that the results of Schirra's medical tests after his recovery also proved highly satisfactory.

"Schirra's pre-flight examinations were similar to those for his predecessors. There were some changes to the blood pressure measuring system. These were principally to the positions of the electrodes and the adhesives used, to provide better readings and to reduce skin irritations. There were also changes to the gain settings in the controller for this system and this astronaut." Special studies were also carried out in order to obtain information on

Everything was explained methodically, but with an occasional dash of humor. (Photo: NASA)

selected body functions and sensations. "These produced biochemical and plasma enzyme determinations and three special measurements: a modified caloric test, radiation dosimetry, and retinal photography. Results of the retinal photography and the modified caloric tests showed no significant changes from prior to the flight. The dosimeters were located in the helmet and underwear and established that the radiation dose posed no hazard. The post-flight analysis of the plasma enzyme studies suggested that the elevations in some parameters were due to muscular activity rather than visceral pooling of the blood. A comparison of the MA-8 biochemical results showed that the astronaut's 9-hour exposure to weightlessness caused no biomedical changes that hadn't been noted after previous manned orbital flights. There were no medical reasons not to embark upon a longer mission."[5]

WELCOMING A HOMETOWN HERO

On Monday, 15 October 1962, Wally Schirra was welcomed back to his birthplace of Hackensack, New Jersey, and to the town of Oradell where he grew up. The evening before a huge and heartfelt post-flight parade was scheduled to take place, a waving crowd of more than 2,000 people was on hand at Newark Airport when the Delta Air Lines plane carrying the astronaut, his parents and his family touched down. As they stepped down from the commercial jet onto the tarmac they were greeted by a 12-feet long banner "Welcome Home, Wally" and the cheers and shouted congratulations of the local citizens.

The dignitaries at the airport to meet the astronaut were led by Governor and Mrs. Richard Hughes, Senators Harrison Williams and Clifford Case, Representative Peter Rodino, and Newark Mayor Hugh Addonizio.

After his 15-minute welcome at the airport, Schirra and his family traveled in an open convertible at the head of a twelve-car motorcade along darkening streets to the Oradell home of the Mayor and Mrs. Frederick Wendel. The couple had obligingly moved out of their house for the night in order to make room for the Schirra family.[6]

At nearby Hackensack, where Schirra had been born in the town's hospital in 1923, the parade next day began at 9:00 a.m. outside the Bergen County Courthouse. Excited crowds of people went into a frenzy of cheering and applause as the fifteen-car motorcade set off. Schirra sat with his wife and children in the rear of an open red convertible for the eight-mile journey, which was due to end at the River Dell Regional High School Stadium in Oradell. NASA Administrator James Webb was in one of the following cars, seated alongside Schirra's beaming parents.

An estimated 40,000 people lined the route to witness the motorcade's procession through Hackensack and River Edge as it made a slower than expected pace through to Oradell. On several occasions the excited, surging throng threatened to get out of hand and it broke through police lines half a dozen times. The lead car, with Schirra and his family smiling and waving at everyone, was forced to stop a number of times as people pressed towards the vehicle in their joyful enthusiasm.

At Newark Airport, ready for the big homecoming day ahead. (Photo: NASA)

A huge crowd had gathered outside the Bergen County Courthouse to see the motorcade set off. (Photo: NASA)

Cheering crowds ensured a slow passage through the streets of Hackensack. (Photo: NASA)

The Schirras could barely believe the huge crowds that had turned out to see the motorcade. (Photo: NASA)

Oradell pulled out all stops to welcome back their hometown hero. (Photo: NASA)

In the normally quiet small town of Oradell, some 30,000 people had congregated to shout and cheer themselves hoarse with pride during the town's tumultuous "welcome home" celebration for their local hero. Red, white and blue bunting and streamers were hung everywhere, and the American flag was proudly displayed on most houses, stores,

Everyone, it seemed, wanted to get into the spirit of the day. (Photo: Bergen County)

Every vantage spot along the route was taken and home-drawn signs were everywhere. (Photos: NASA)

Hometown pride was evident all over town. (Photo: NASA)

and public buildings. "Welcome, Wally" signs were in abundance. Here too the police were kept busy continually holding back the eager crowds of men, women and children, who were out celebrating what the town called "Hurrah Schirra" day. The local schools had closed to allow everyone possible to join in the festivities.

The motorcade moved through River Edge and the lower part of Oradell, and then swung through New Milford before re-entering Oradell. The route took Schirra by the house at 371 Maple Street where he had once lived. He pointed it out to his family as they passed by, as well as the church he had attended.

According to Oradell borough historian George Carter, "The parade came up Oradell Avenue and turned left at the railroad station onto Maple Avenue. They drove Wally by his house, then turned right up Ridgewood Road to Kinderkamack Road. Another right-hand turn and on down Kinderkamack Road past the reviewing stand. Then left up Oradell Avenue, left onto Prospect Avenue, and on to the football field at the senior high school."[7]

At the football stadium, Schirra was overwhelmed to see that around 8,000 people had gathered to see him, cheering and chanting over and over again, "Hurrah, Schirra! Hurrah, Schirra!" As his limousine pulled up, the school band and students serenaded him with a revised version of the song *When Johnny Comes Marching Home*, with the lyrics changed to "When Wally comes flying home again, hurrah, Schirra! We'll give him a hearty welcome then, hurrah, Schirra!" He later remarked that the greeting left him "much more shattered than I was at the countdown."[8]

A cheerful Schirra acknowledges all the waves and shouts as the motorcade passes through Oradell. (Photo: NASA)

Everyone was hoping for a glimpse of the hometown astronaut. (Photo: NASA)

The house where Wally once lived was also dressed up for the occasion. (Photo: NASA)

The Dwight Morrow school band was at the stadium to greet him. (Photo: NASA)

The newly erected road marker to honor Schirra's accomplishment. (Photo: NASA)

During his appearance at the stadium, Schirra assisted in the dedication of Schirra Park in the center of town, near the New Jersey Transit railway station. A road marker was also unveiled in honor of his historic space flight, erected on the nearby corner of the town's main commercial thoroughfare, Kinderkamack Road, and Oradell Avenue. The sign read:

ORADELL
Home of Commander Walter M. Schirra, Jr., USN,
The First Jerseyman to Orbit the Earth, Oct. 3, 1962

RECEIVING AWARDS

After the early formalities and dedications had been completed, NASA Administrator James Webb presented Schirra with the agency's Distinguished Achievement Award. Prior to presenting the award, Webb told the crowd, "The strength of America comes from places like Oradell – they produce men like Commander Schirra." The crowd erupted into loud cheers.

Part of the massive crowd that had gathered at the football stadium. (Photo: NASA)

The Schirra family seem a little overwhelmed by the huge reception. (Photo: NASA)

Webb then read out a message of congratulations received from President Kennedy, in which he described the MA-8 flight an "exceptional achievement." The president's message continued:

> I would like to add my personal congratulations. The entire Mercury team is to be commended for its dedicated efforts to develop U.S. capabilities in space – to get our space program moving from the position of second best to one of world leadership. That we are climbing back up the ladder is apparent in the matchless performance of our recent flight. I am convinced that although we are still second in hardware, we bow to no one in the quality of our space team – that their devoted and determined efforts will, in this decade, restore our leadership.
>
> I am particularly proud of Commander Schirra for the great professional skill and personal courage he demonstrated in his magnificent flight.

In his response, Schirra cited Dwight Morrow High School, the Newark College of Engineering, the U.S. Naval Academy, and Oradell as contributing to his development, and called his flight "the sum effort of many."

Schirra receives NASA's Distinguished Service Medal from James E. Webb, the agency's administrator. (Photo: NASA)

Following this, Governor Richard Hughes congratulated Schirra on behalf of his native state of New Jersey, hailing him for his "pioneer spirit of bravery and courage combined with the trained intelligence of modern man" in his conquest of space. Dr. Robert van Houten, president of the Newark College of Engineering, which Schirra attended for three semesters before going to the U.S. Naval Academy, presented the astronaut with the college's Edward F. Weston Distinguished Alumnus Award.

The Schirra family also received as a gift an 88-piece china service embossed with the *Sigma 7* logo, a silver tray from the people of Oradell with the inscription "the first Jerseyman to orbit the Earth," and a black-and-white screen print by Lithuanian-born New Jersey artist Ben Shahn.[9]

Once the ceremonies had ended, everyone in the official party filed back into their automobiles to journey to the Hackensack Golf Club, which hosted a private luncheon attended by 252 guests at which there were no speeches made.

Some years later, Oradell would further honor its hometown astronaut by renaming one of the main thoroughfares in the exclusive Blauvelt area as Schirra Drive.

That afternoon, at a special ceremony held in Oradell's Schirra Park, the astronaut was presented with a specially-made merit badge by his old scout troop, No. 36, for "achievements in Astronautics." In handing the badge to Schirra, 15-year-old Eagle Scout Ronald Nillsen called it "the first Astronaut Merit Badge." He then explained that the requirements were:

- Present yourself fully clothed for the test, a nine-hour space journey in the vicinity of the Earth
- Take off and come to an altitude of at least 100 miles
- Maintain minimum speed of not less than 16,000 miles per hour
- Make no fewer than six orbits, coursing a distance of at least 155,000 miles
- Upon landing avoid carelessly placed rescue vessels.

Wally laughed loudly at the final requirement, and responded by telling the scouts, "I hope many of you will have the opportunity to win this badge in the years to come."

To wrap up what had been a hectic but special day for Wally Schirra and his family, they revisited Mayor Wendell's home prior to setting off for New Jersey's Teterboro Airport along the same nine-mile route which he used to cycle along from his home in order to watch the airplanes take off and land. Their flight left at 5:00 p.m., bound for Washington, D.C.

VISIT WITH THE PRESIDENT

The following day, 16 October, the Schirra family was granted their audience with President Kennedy in the Office of his Naval Aide in the White House. Sitting in his rocking chair, the president asked Schirra some penetrating questions relating to his space flight, and was especially interested in his extended period of weightlessness. "It's obvious from his questions that he does a lot of homework," Schirra later told reporters.

A delighted Wally Schirra receives his special merit badge from members of his old boy scout troop. (Photo: NASA)

President Kennedy talks with the Schirra family in the White House. (Photo: John F. Kennedy Library and Museum)

As Jo Schirra looked on quietly, President Kennedy also engaged in small talk with the two children, asking young Walter where he went to school. At one stage, Suzanne looked shyly at the president and said, much to his delight, "I know who you are." He asked Suzanne how old she was and she held up five fingers. When asked if she had watched her father's flight on television, she gave a little nod. The president asked the children if they would like to see the ponies kept at the White House for his children, and when they agreed he took them all to meet the ponies, one of which was Caroline Kennedy's horse, Macaroni. As parting gifts, 12-year-old Walter received a tie clasp from the president and Suzanne was presented with a charm bracelet.

It was an unusually brief informal meeting, lasting just 18 minutes from when the Schirra's first sat down at 9:30 a.m. They were unaware that the president had been awakened earlier that day with the dire news that a missile launch site and two new military encampments had been detected in west central Cuba, although the missiles were not yet operational. As a result, Kennedy had ordered seventeen key military, political, and diplomatic advisers to assemble at the White House at 11:50 a.m.

A happy, relaxed family after their meeting with the president. (Photo: NASA)

After the president had presented Schirra with the Distinguished Service Medal, it was time to make their farewells. Once outside, Schirra talked with newsmen, saying that he expected to return to Houston that night. He would then fly to Cape Canaveral the next day, where he had to complete a written report on his nine-hour space flight and turn his attention to design problems with the new, two-seater Gemini spacecraft. He was asked if he felt that he needed a vacation. With a smile, Schirra responded, "I don't think I need any now. I had mine while I was flying."

He then delivered a talk to a crowd outside of NASA's headquarters building, after which he was swamped with people wanting to shake his hand and get his autograph.

Outside NASA headquarters, Schirra gave a short speech to an appreciative crowd. (Photo: NASA)

Later that day, Schirra, wearing his full dress uniform, attended a ceremony at the Pentagon where Secretary of the U.S. Navy Fred Korth pinned the Navy's astronaut wings on his jacket as Jo looked proudly on. Meanwhile, young Walter and Suzanne waited patiently on the steps outside and were photographed comparing the gifts they had received from the president.

Wally Schirra signs a number of autographs for NASA employees. (Photos: NASA)

A proud Captain Wally Schirra receives his astronaut wings from Secretary of the U.S. Navy, Fred Korth, as his wife Jo looks on. (Photos: NASA)

REFERENCES

1. Paul Haney email correspondence with Colin Burgess, 30 September 2003
2. NASA *Space News Roundup*, "Text-Book Flight Near Perfect," NASA MSC, issue 17 October 1962, pg. 2
3. *Evening Independent* newspaper (St. Petersburg, FL), unaccredited article, "Schirra Ended 'Textbook Flight' Reluctantly," issue 8 October 1962, pg. 5-A
4. NASA *Space News Roundup*, "Text-Book Flight Near Perfect," NASA MSC, issue 17 October 1962, pg. 2
5. Manfred (Dutch) von Ehrenfried, *The Birth of NASA*, Springer-Praxis Publications, New York, NY, 2016
6. The *Spokesman-Review* (Spokane, Wash.) newspaper, unaccredited article, "Cheering Crowds Welcome Schirra," issue Saturday, 16 October 1962, pg. 21
7. George Carter email messages to Colin Burgess, 24–30 June 2015
8. NASA Space Flight News, "Schirra Gets New Jersey's Welcome; NASA Award and Chat With JFK," MSC Houston, TX, issue 31 October 1962, Pg. 2
9. The *Spokesman-Review* (Spokane, Wash.) newspaper, unaccredited article, "Cheering Crowds Welcome Schirra," issue Saturday, 16 October 1962, pg. 21

8

Mercury, Gemini and Apollo veteran

As far as NASA was concerned, Project Mercury concluded with the successful (albeit somewhat trouble-plagued) 22-orbit flight of Gordon Cooper aboard spacecraft *Faith 7* on 15/16 May 1963. There were no more single-pilot flights manifested, although for a time Alan Shepard led a spirited but ultimately unsuccessful campaign to add one final Mercury flight, which he obviously wanted to fly. He would later fall victim to an ear ailment which forced him to stand down from the flight rotation for several frustrating years.

Nevertheless, Project Gemini moved ahead. Schirra, along with the other Mercury astronauts, was ready and eager to fly a Gemini mission and assist in the development of the program. But they had rivals following the selection of the nine new astronauts who were in many ways better pilots with more educational qualifications. Naturally, everyone wanted to be selected to command the best available missions, with the first Gemini flight and the first rendezvous being regarded as plum test pilot assignments.

Schirra was in the rotation and stood ready to command his own mission. Shepard and Slayton had fallen off the roster because of medical problems. A dispirited Glenn came to the sad realization that he might not fly again due to his enormous popularity and PR value, and quit the space agency. Carpenter was still under a cloud following his flight and would temporarily leave NASA for an assignment with the U.S. Navy's SEALAB venture. Having flown the last Mercury mission, Cooper had to accept that he might be assigned a later Gemini mission that no one else really wanted. This had left Grissom and Schirra with the best options. Grissom had the edge because he had done more work on the development of the new two-seater spacecraft.

On 13 April 1964 the space agency named Gus Grissom and Lt. Cdr. John Young as the commander and pilot respectively of the first manned Gemini mission, which was then scheduled to complete three orbits of the Earth early the following year. Grissom would become the first person to make two flights into space. Their backup team was Wally Schirra and Maj. Thomas Stafford, who were expected to "rotate" to fly a later mission.

C. Burgess, *Sigma 7*, Springer Praxis Books, DOI 10.1007/978-3-319-27983-1_8

ON TO GEMINI

Project Gemini was developed as an intermediary bridge between the Mercury and Apollo programs, with the objective of testing equipment and mission procedures in Earth orbit to assist Apollo planners. As outlined by NASA, the goals included long duration flights in excess of the requirements of a lunar landing mission; rendezvous and docking of vehicles in Earth orbit; onboard orbital navigation; the development of operational proficiency of both flight and ground crews; the conduct of experiments in space; extravehicular operations; and active control of the re-entry flight path in order to achieve a precise landing point.

The prime and backup crews for Gemini 3 looking extremely self-conscious in their space suits and helmets. (Photo: NASA)

On 23 March 1965, Grissom and Young were launched for the Gemini 3 (GT-3) mission that flew three times around the world before splashing down in the Atlantic. The flight itself was hailed as a great success and a fitting prelude to the missions to follow. However, the crew would have to live down an incident involving a corned beef sandwich – with a certain Wally Schirra spotlighted as the culprit.

Gemini 3 was slated as a flight of less than five hours covering three orbits, and while the two men might not have suffered hunger pangs during that time, one of their tasks was to test a number of plastic food tubes that were packed with some newly developed dehydrated food items. These could be reconstituted by squirting liquid into the tubes using a water gun. Scientists were eager to know for future flights how the packaging fared and how well the crew could work and eat at the same time.

Just under two hours into the flight Young decided it was time to tackle the food task while Grissom concentrated on flying the spacecraft that he had unofficially nicknamed *Molly Brown*. Then, to Grissom's surprise, Young said, "Would you care for a corned beef sandwich, skipper?"

Grissom thought his pilot was joking, but when he looked he found that Young was holding out a true-to-life corned beef on rye sandwich, which it was later revealed came from Wolfie's Delicatessen in nearby Cocoa Beach.

"Where did that come from?" Grissom asked.

"I brought it with me," Young replied. "Let's see how it tastes. Smells, doesn't it?"

"Yes, [and] it's breaking up," Grissom responded, as each of them took a quick bite of the delicious sandwich. "I'm going to stick it in my pocket."

"It was a thought, anyway," the normally dour Young said.

"Yep," Grissom replied, and said no more about it.

NASA suit technician Al Rochford recently revealed to the author just how this item came to be on board. "Gus and John were in the suit room that was next to their living quarters. Wally walked in with this corned beef sandwich that was wrapped in grease-soaked napkins and gave it to me to put in John's lower suit pocket. Well I told Wally, 'I will have to clean this up,' so I slipped off to a back room and cleaned up the grease, wrapped it in clean paper towels and placed it in a zip-locked bag, then put it in John's left lower pocket without anybody seeing it. I looked up at John as I was putting in to his pocket. He saw me but said nothing."[1]

Both astronauts later suffered some mild rebuke from their higher-ups after the flight for this little antic, as they had violated the rules which said they were to eat nothing but the dehydrated food, and because the floating crumbs from the sandwich could possibly have caused something to go wrong. Schirra naturally thought it was all great fun. He knew that Grissom constantly complained about the "dehydrated delicacies concocted by NASA nutritionists," and so he had decided that the first Gemini crew would have a little something extra to chew on while in orbit.

But that was not the end of the "corned beef affair." The U.S. Congress, which was responsible for NASA's budget, was far from impressed with the notion that astronauts felt a practical joke was appropriate behavior on an expensive and high-profile mission. Schirra, Grissom, and especially Young, received a full dressing-down from Congress and the press, despite senior technical management coming to their defense. In fact the

Young and Grissom suit technician Joe Schmitt (partially obscured at *left*) prior to their GT-3 mission. (Photo: NASA)

Associate Administrator of the Office of Manned Spaceflight was quick to tell Congress that, "there was no detriment to the experimental program that was carried on, nor was there any detriment to the actual carrying out of the mission because of the ingestion of the sandwich."

But NASA Administrator James Webb sided with Congress, and Young received an official reprimand. Fortunately this had no lasting impact on his astronaut career, as in 1972 he walked on the Moon as commander of Apollo 16 and in 1981 he commanded the maiden mission of the innovative space shuttle.

CALLING OFF A CHASE

Under the system that was then in place, Schirra and Stafford would rotate to primary crew status three flights later. This became official on 5 April when they were named as the prime crew for the fourth manned Gemini mission, which was slated as the first rendezvous and docking flight in the program. After Gemini 3, NASA had decided to change the mission numbering from Arabic to Roman numerals, therefore their flight would be Gemini VI.

The Gemini VI crew of Wally Schirra and Tom Stafford. (Photo: NASA)

On 3 June 1965, the Gemini IV mission launched Jim McDivitt and Ed White into Earth orbit with the undoubted highlight being the first American spacewalk (EVA or Extra Vehicular Activity in the NASA lexicon) carried out by White. Tethered to the spacecraft, he floated freely for 20 minutes before reluctantly returning inside. But he wasn't the first to accomplish this feat because Soviet cosmonaut Alexei Leonov had done so in March. Two months after Gemini IV, on 21 August, Gordon Cooper and Pete Conrad were launched aboard Gemini V on an eight-day mission to simulate a return flight to the Moon. They were plagued by a faulty fuel cell designed to provide electrical power, but managed to achieve their assigned flight duration. The next step was to conduct a rendezvous and docking in orbit during a two-day mission. The man who would command that crucial flight was Wally Schirra. The launch was planned for 25 October.

The plan called for an unmanned 6-meter-long Agena target vehicle to be launched by an Atlas rocket from Cape Canaveral on the morning of 25 October, followed 101 minutes later, as the Agena completed its first revolution, by Gemini VI. The Agena had a docking cone at its forward end into which the spacecraft could be inserted and held with a trio of docking latches. The rendezvous was planned to take place on the fourth orbit, between Australia and Hawaii, following which there would be several practice dockings and separations. It would be a crucial flight and Schirra said at the time, "If we can't rendezvous and dock, we're stalled on the Moon trip." But then in typical Schirra fashion he said the rendezvous should be "a piece of cake."

The Atlas lifted off right on schedule at 10:00 a.m. on 25 October, with Schirra and Stafford already aboard their spacecraft on Launch Complex 19 (LC-19), some 6,000 feet away. The Atlas performed perfectly during the five minutes of the ascent, then it separated, leaving the Agena to continue into orbit. But then things went dramatically wrong, with Mission Control reporting "a dramatic loss of telemetry" just as the Agena attempted to ignite its engine. It was later determined that the Agena had exploded.

At 10:54 a.m. the Gemini VI mission was postponed indefinitely. Because there would have to be a thorough investigation into the loss of the Agena vehicle, NASA decided to switch its attention to Gemini VII. However, very soon a whole new and bold plan emerged, as everyone regrouped.

The mission profile for the Gemini VI rendezvous and link-up with the Agena target vehicle. (Photo: NASA)

Atlas-Agena target vehicle GATV-5002 lifts off from Launch Complex 14 ahead of the Gemini VI rendezvous mission, but was lost in an explosion before reaching orbit. (Photo: NASA)

The chief instigators were Walter Burke and John Yardley of McDonnell, the chief project officers for Mercury and Gemini, who proposed the "salvo launching" of two manned Gemini missions. At that time, Gemini VII and its crew of Frank Borman and James Lovell was scheduled to be launched on a two-week mission in early December. Burke and Yardley explained that if Gemini VI could be launched while Gemini VII was still in orbit, Schirra and Stafford could rendezvous with Gemini VII as the target. "Frank Borman was as enthusiastic as I was," Schirra later recalled, "but he said 'no' to attaching a docking adapter to his spacecraft. We would rendezvous but not dock, if the idea was approved."[2]

Initially the feeling was that it couldn't be done, because NASA had never tried to launch two manned spacecraft in such a short period, and the network of tracking and communications systems was not designed to cope with multiple vehicles. But Burke and Yardley pressed the idea with Robert Gilruth, in charge of the Manned Spacecraft Center, who eventually gave the plan his support. In turn, he discussed the idea with his Flight Director Chris Kraft, who was initially skeptical but then he too saw how it could be managed and he also gave his support the dual flight. At a press conference given at his Texas ranch on 28 October, President Johnson announced the news of this audacious new flight plan involving the two Gemini spacecraft.

THE MOST DANGEROUS MOMENTS

A plan was mooted for Stafford to undertake an EVA from Gemini VI at the end of a metal-plated lifeline and propel himself 100 feet to Gemini VII, possibly to exchange places with Jim Lovell. But in November, when NASA announced that the target date for launching Gemini VII was set for 4 December, it also ruled out any possibility of a spacewalk and crew swap being added to the dual flight.

Gemini VII lifted off from LC-19 as planned and entered the desired orbit to start its 14-day space marathon. The Titan II which would launch Gemini VI-A (as the mission had been redesignated) was moved out to the pad with a launch set for eight days later, on Sunday, 12 December. This time, things did not go quite as smoothly.

It was 9:54 a.m. EST on launch day, at the end of a faultless countdown. As Schirra and Stafford prepared for lift-off, each man procedurally gripped a D-ring handle (his "chicken switch") ready to eject from the spacecraft and parachute to the ground a safe distance away if there was imminent danger of an explosion.

Finally, the count reached zero. Both men heard the Titan pre-valves open and the turbopumps whir into instant life. A burst of reddish-brown smoke erupted at the foot of the 90-foot-tall rocket, billowing away as the engines ran up to their rated 430,000 pounds of thrust. But less than two seconds later, as millions of Americans watched live on television, the engines shut down and the Titan remained locked firmly on the launch pad, held down by four huge bolts, and briefly engulfed by billowing smoke. "Shutdown Gemini Six!" was immediately relayed to the crew from Mission Control.

There were strict mission rules covering this scenario. Because the onboard clock had started, those rules stipulated that Schirra, as commander, was to pull the D-ring beneath his seat to activate both ejection seats and blast them away from what could momentarily

The launch pad shutdown of the Gemini VI-A launch. (Photo: NASA)

become an explosive holocaust. But Schirra was not one given to panic. He had ridden a rocket before and couldn't sense any motion to indicate that a lift-off had occurred, and while he was still prepared to initiate an ejection he decided to stay put.

Schirra and Stafford only had moments to decide what to do. Their eyes swept the instrument panel for clues. If the fuel and oxidizer pressures were rising, an explosion would likely follow. But 15 seconds after shutdown, Schirra calmly announced, "Fuel pressure is lowering." With each passing second the danger diminished.

"A light came on in the spacecraft saying we had liftoff," Schirra later recalled. "I heard from the blockhouse that the clock had started, which means we had lifted off. But I knew we hadn't lifted off. It was a gut feeling. Stafford didn't know what was going on but I had the experience of a Mercury flight and my butt told me we hadn't left the pad."[3]

Although Schirra should have initiated an ejection, he knew "in my soul, in my body, that we had not lifted off." Tom Stafford also remained calm and said, "O.K., Wally, I buy it." For the next few moments the two men simply sat there, tense, and breathing heavily.

Meanwhile 185 miles overhead and on its 118th orbit, Gemini VII was flying over the Cape, with Borman and Lovell hoping to witness the launch. But as they watched they saw the lift-off aborted. "We saw it ignite. We saw it shut down," Borman later reported.

According to Gene Kranz, the Assistant Flight Director in Houston, "Schirra and Stafford basically decided that they were not going to eject and they held their fire. I had a controller in Mission Control who had similar responsibilities. He had his finger on the abort switch and he also decided to hold fire. So this is probably one of the most short-term, riskiest, one hundred percent correct decisions that we ever made."[4]

About 40 minutes after the launch shutdown, the booster was declared safe and the service tower was rolled back in to envelop the Gemini spacecraft. Nearly an hour after the dramatic events had unfolded, Schirra and Stafford were finally assisted out of their couches. Their extraordinary courage in not ejecting would later be rightfully hailed as heroic. Their coolness under extreme pressure had saved the mission and averted what might have been a lengthy delay to the entire Gemini program.

After watching the abort on a color TV at his ranch President Johnson told reporters, "Our disappointment is exceeded by our pride in astronauts Walter Schirra and Thomas Stafford and flight directors of NASA. They acted with remarkable courage in the face of danger and potential disaster."

An initial analysis of the telemetry identified the fault as an electrical plug designed to carry data from the malfunction detection system to the nearby launch blockhouse. The plastic plug, worth about two cents, was meant to be pulled out of the base of the vehicle as it lifted off, but it had fallen out of its own accord two seconds early. It was the disconnection of this plug which started the onboard clock. It was further reasoned that when the inadvertent disconnection was seen by the malfunction detection system this had automatically shut down the engines, which had been burning for 1.2 seconds.

However, engineers routinely examining other telemetry discovered the engines had stalled *prior to* the plug disconnecting. A physical inspection found a dust cover which had been placed in the oxidiser inlet of the gas generator by the manufacturer during engine assembly and never removed.

The next day, an Air Force safety review board held a 90-minute conference which conditionally cleared the troubled Gemini VI-A mission for a third launch attempt three days later.

Gene Cernan, later to achieve lasting fame on the Apollo 17 mission as the last man to walk on the Moon, witnessed the launch abort with some fellow astronauts and later described Schirra as a "cold-nerved pilot" in the potentially calamitous moments which followed ignition. "[The] Titan's massive engines ignited, and as the smoke and flame belched around the pad, the damned rocket didn't lift off! Wally and Tom sat through the firestorm, perched on the edge of disaster while the rocket bellowed and shook around them. We watched in awe, knowing that the book called for the crew to bail out of there before the rocket blew up under them. Schirra gripped the orange D-ring that would eject both him and Stafford from the endangered spacecraft, but he didn't pull it! Schirra had balls. He broke all the rules that day, but saved the program from a potential show-stopping disaster. The hot engines shut down, the raging Titan calmed, then Wally and Tom climbed out of the spacecraft and shrugged off the entire episode."[5]

SPACE CHASE

Right on schedule at 8:37 a.m. on 15 December, Gemini VI-A was finally launched. It entered orbit 6 minutes and 20 seconds later. At that time Gemini VII was entering its 162nd revolution of the planet and was 270 miles downrange of Cape Canaveral. The flight plan was for Schirra and Stafford to carry out a series of maneuvers designed

Gemini VI-A is launched into the Florida skies to start the space chase for Gemini VII. (Photo: NASA)

to culminate on the fourth orbit in a rendezvous with the Gemini VII spacecraft, having covered a total distance of 103,000 miles.

Just under five hours later, at 2:28 p.m. EST, the United States achieved its history-making rendezvous with the two spacecraft traveling nose-to-nose, several feet apart, 183 miles above the Earth. In an extraordinarily precise piece of flying by Schirra he performed four orbital changes prior to the final delicate maneuvering that consumed only half of the allocated fuel and ended with the two spacecraft flying in formation. He reported that in his opinion there would be no problem rendezvousing and docking with another vehicle in space. "It's easier than in the Gemini docking trainer," was his laconic remark.

Flight Director Chris Kraft and his flight control team were jubilant. "They were about as close as they could get without actually touching," he later stated.

Altogether, the two spacecraft spent five and a half hours orbiting together, often punctuated by humorous exchanges between the two crews. Schirra would maneuver Gemini VI-A back and forth and up and down in relation to the other craft, while both crews snapped some of the most iconic photographs seen to that time from any space mission.

Having completed the rendezvous program, both crews settled down for the night, ready to catch up on some sleep after a truly momentous day's activities. Chris Kraft summed everything up when he said, "It was a great day."

As Gemini VI-A was preparing to return to Earth the next morning, it sent a scare through Mission Control when Stafford reported sighting a "strange object" on a low-trajectory polar orbit. But the tension was broken moments later when Schirra gave a rendition of *Jingle Bells* on a tiny mouth organ, accompanied by Stafford on a set of tinkling bells that they had brought with them for the occasion. The "strange object" was a joke sighting of Santa Claus making his Christmas run. "Wally came up with the idea," Stafford told *Smithsonian* magazine in 2005. "He could play the harmonica, and we practiced two or three times before we took off, but of course we didn't tell the guys on the ground. We never considered singing, since I couldn't carry a tune in a bushel basket."[6]

After nearly 26 hours in orbit, Gemini VI-A splashed down in the North Atlantic about 600 miles north of Bermuda, and within 11 miles of the recovery carrier USS *Wasp* (CVS-18). Thus a mission which had been beset with major problems, ended gloriously as a result of the remarkable courage and skill of the crew and the can-do efforts of everyone associated with the flight.

Wally Schirra (now promoted to the rank of captain) and Frank Borman, together with their wives, were later sent on a goodwill tour around the world that took in such countries as Japan, South Korea, Taiwan, the Philippines, and Australia.

"But all in all Gemini was a success, a totally satisfying experience," Schirra later recorded. "Gemini would fade into history as we went to the Moon in Apollo – the price of success!"[7]

Gemini VI-A approaches Gemini VII for the historic rendezvous, 15 December 1965. (Photo: NASA)

Gemini VI-A closes to within 35 feet of its sister ship. (Photo: NASA)

Safely aboard the carrier USS *Wasp*, Stafford and Schirra celebrate the end of their epic flight. (Photo: NASA)

TRAGEDY, AND A TROUBLED TIME

On 29 September 1966, Deke Slayton officially announced the selection of the first two crews to be assigned Earth-orbiting Apollo test flights. Gus Grissom was to command the first mission, which was initially designated Apollo-Saturn 204 (AS-204) and then renamed Apollo 1. His crew would be Ed White from Gemini IV and rookie astronaut Roger Chaffee. The second test flight fell to Wally Schirra, along with a pair of rookie astronauts: Donn Eisele and Walt Cunningham.

"It wasn't a crew I planned to use on lunar landing missions," Slayton wrote in his 1994 memoir *Deke!* "Wally was making noises about retiring, and I figured to move Donn and Walt over to Apollo Applications. The mission wouldn't break any new ground. As the second and last flight of a Block I command module it wouldn't even be equipped for rendezvous and docking with a lunar module. It was on the schedule [simply] to give us a chance to pick up whatever might get missed on the first manned Apollo flight."[8]

Despite being assigned to another mission, an important milestone in itself, Schirra was far from happy with the role they had been handed and insisted on changes being made, as he later recalled. "Our flight was to be identical to that of Grissom, White and Chaffee – ten to fourteen days in Earth orbit in a Block I spacecraft. I argued it made no sense to do a repeat performance, and I succeeded in getting the mission scrubbed. I also got my crew eliminated from the early rotation. I had hoped to have us assigned to the first Block II flight, but it went to Jim McDivitt, Dave Scott and Rusty Schweickart. We replaced the McDivitt crew as backup to the Grissom crew. And [as events finally turned out] the third Apollo mission, the first to be launched by a Saturn V, was given to a crew commanded by Frank Borman."[9]

Training for the named crews continued through 1966 and the early part of 1967, with the first mission set for 21 February. But then, on 27 January, NASA's worst nightmare came true when the Apollo 1 crew died in a fire which swept through the oxygen-soaked interior of their spacecraft during a "plugs-out" test atop an unfueled Saturn IB rocket on Launch Complex 34. With their capsule over pressurized by the fire, the astronauts were unable to open the complex inward-opening hatch and died within seconds from cardiac arrest as a result of inhaling toxic fumes.

What resulted from this launch pad catastrophe was a rebuilt and much safer Apollo spacecraft, much better procedures, and a much improved culture within NASA and its contractors.

Once sufficient progress had been made, NASA announced that the backup crew of Schirra, Cunningham and Eisele would fly a revised version of the orbital shakedown which had been planned for Apollo 1. Coming after a series of unmanned tests of the Saturn V launch vehicle and the lunar module, Schirra's mission would be designated Apollo 7.

Still getting over the loss of Grissom's crew, Schirra threw himself wholeheartedly into training with a determination to do the best possible job before he quit the space agency.

The ill-fated crew of Apollo1: Gus Grissom, Ed White and Roger Chaffee. (Photo: NASA)

The Apollo 7 crew of Donn Eisele, Wally Schirra and Walt Cunningham. (Photo: NASA)

> Apollo 7 was going to be my last mission. I intended to retire, and I would make it official before we flew. There were a number of reasons. For one I wanted to quit while I was ahead. I also wanted it to be clear that I was single-minded about the Apollo 7 mission, that I cared about nothing else.
>
> I had changed over the span of time that encompassed my three flights. As the space program had matured, so had I; I was no longer the boy in scarf and goggles, the jolly Wally of space age lore. When the original crew of the first Apollo was lost, I became deeply involved in deciding where we were headed. And when I realized we would try again with me in command, I resolved that the mission would not be jeopardized by the influence of special interests – scientific, political, whatever. I was annoyed by people who did not consider the total objective of the mission. I would not be an affable fellow when it came to decisions that affected the safety of myself and my two mates.[10]

Launch pad leader Guenter Wendt noticed an immediate shift in Schirra's attitude. "After the fire he changed quite a bit. He was always a happy-go-lucky guy, but now he knew he had to do better than that."[11]

In his memoir *Schirra's Space*, he recounted an incident at the Downey, California plant when he was with some NASA dignitaries inspecting the Apollo spacecraft that he would fly. He was dressed in pure white protective clothing and cap, and when he stepped into the spacecraft he was ultra-careful not to tread on, and possibly damage, any electrical or mechanical parts. Unfortunately it was a tight squeeze, and his knee accidentally brushed up against a bundle of wires. "When it did, I felt a sharp slap on my face, and I heard a woman's voice: 'Don't you dare touch those wires. Don't you know we lost three men?' When she was told who I was, she felt embarrassed, but I assured her she needn't be. 'Keep it up,' I said. 'I want people like you working on this spacecraft.'"[12]

Schirra was adamant there must be no compromises on the issue of fire safety. He even took to the extreme of quitting his pack-a-day cigarette habit, but he would find some fire prevention methods to be quite ludicrous, such as an on-board ban on books or magazines, and any objects made of paper, including a pack of cards. Schirra, who had an intense dislike of long space flights, thought things were getting just a little out of hand, arguing that the crew had to do something to relieve the tedium. They could not just stare out of the windows for 11 days.

There were other less-pressing issues with the flight that had to be straightened out before launch. At one stage mission planners, physicians and psychologists informed Schirra that coffee would not be loaded in the spacecraft, saying that it had no calorific value and was merely a stimulant. At this, Schirra blew up. "Yeah, all it does is make me happy!" he responded. Did they seriously expect a Navy man to go without coffee for nearly two weeks? He knew that the fuel cell that supplied electrical power for the spacecraft produced hot water as a by-product at around 155 degrees Fahrenheit. The water could be squirted into a bag containing freeze-dried coffee granules. It was an argument he was determined not to lose, and so he carried out one of his famous ruses in order to get his way. During a prearranged break at an important meeting of senior NASA management figures, Schirra arranged to have a cart containing Danish pastries wheeled into the room. Everyone soon noticed something was missing, at which time he stood up to address the meeting.

"Gentlemen," he said, "since you deem it inappropriate for the crew of Apollo 7 to drink coffee on the mission, I thought you might try doing without it for just one day!" He won the argument and coffee was subsequently loaded for the flight.[13]

THIRD AND LAST FLIGHT

On 20 September 1968, Schirra officially announced that Apollo 7 would be his last space flight and he would then retire from NASA and the U.S. Navy, effective 1 July 1969. "I don't think I'd have the steam left to do another one," he told reporters in making his announcement.

The following month, on the morning of Friday, 22 October, the crew of Apollo 7 was tightly strapped inside Command Service Module No. 101, ready to complete the mission for which their three colleagues had given their lives 21 months earlier. Many crucial lessons had been learned in that time, and Schirra had proved to be a true and determined taskmaster, totally unaccepting of laxity and bad practices, and keeping a stern eye on every system, instrument, and fixture as their spacecraft was assembled and fitted out. Now it was time to put all the hard work to the test and to validate the Apollo CSM as the spacecraft that would one day carry Americans out to the Moon.

Launch Complex 34, where the fatal fire had occurred, was the scene for this latest mission and tensions were high, but there was also confidence because the "old pro," Wally Schirra, was the man in command. He soon proved this by pointedly asking if the winds that were constantly gusting around the launch pad were strong enough for the flight to be delayed. As the winds encroached on safety margins he became more assertive that the situation could prove dangerous during the launch. But everyone – including Schirra – had a great deal of faith in the capabilities of the Saturn IB, and as controllers cautiously monitored the wind strength the countdown finally reached the point of ignition.

Schirra would later say he didn't believe they should have launched with 22 m.p.h. winds that could have blown them back over the beach in the event of an early abort, because the couches in the spacecraft were of an older type that wasn't suitable for a ground impact. "So someone broke that rule. I didn't. I was compromised."[14]

At 11:03 a.m. EST, three minutes after the scheduled lift-off time, a burst of flame erupted beneath the 22-story-tall Saturn IB rocket. Once unleashed, the vehicle rose from the pad on a billowing tail of orange fire and climbed into the blue Florida skies consuming propellant at a rate of 720 gallons per second. At the 55 second mark, the call from Mission Control in Houston was "she is going straight and true."

The first stage separated as planned after two and a half minutes and the single-engine second stage took over. As the 58-foot stage reached its full thrust, Schirra reported, "All beautiful." Then as Apollo 7 entered orbit 8 minutes after leaving the ground he assured controllers, "She's riding like a dream."

Some 3 hours later, near the end of the second orbit and about the same time as a crew would do so on a lunar mission, Apollo 7 executed the first major operation by detonating explosive bolts to separate the spacecraft from the spent stage.

Despite developing a bad head cold on the first day in space, Schirra insisted that they stick to the agreed flight plan. When Mission Control began asking him to move some tests

Lift-off of the Apollo 7 mission; Wally Schirra's final space flight begins. (Photo: NASA)

around within the schedule, he became increasingly testy. His annoyance became obvious to everyone when he was asked to advance a TV transmission to that day instead of the following day. NASA wanted to demonstrate some positive public relations on the first Apollo mission, but Schirra was only interested in the engineering safety aspects of the mission, as he explained for the book *In the Shadow of the Moon*:

> I eventually said, "Okay, if you're going to be violating rules, guess what I'm going to be doing? I am going to judge these rules from now on. If you are going to break that rule and not give me a chance – then I am going to break some of the rules that you have given me problems with." I didn't want to do things that hadn't been tested in their proper sequence. We were to test the circuit of the television set prior to using it. That was scheduled for one day; then the next day we were to turn it on. But the day [on which] they requested us to play games with the television, I was trying to do a rendezvous with the booster. I didn't want to mix that up with something else that was not important. They wanted the television on a particular day, and it wasn't scheduled for that day. I said, "We'll just put it on tomorrow." That made sense to me – but not to them. I was pretty annoyed with Chris Kraft and the flight directors, who were ordering me. I said, "You don't order a commander!" That's why we called the position a commander, after all, not pilot and copilot. The pushing around should only be done by the engines and thrusters, not the flight controllers. But having said all that, I felt a lot for the flight controllers, and worked with them, not against them.[15]

In spite of suffering with his head cold and a lingering resentment at being told to change things, when the live television was relayed to the viewing public he was his usual affable, Jolly Wally self.

Over the next few days the crew performed what could be deemed the perfect test flight for a new vehicle, carried out with the eyes of the world upon them, particularly following the fatal pad fire 21 months earlier. Each of them performed his duties with competence and confidence in what they and their crewmates were doing. Although switching on the TV camera was the cause of much banter between Schirra and the ground, several entertaining live television shows were beamed to the ground from space for what Donn Eisele described as, "The one and only original Apollo orbiting road show." The crew gave millions of Earthbound television viewers a close look at the spacecraft's control panels as it swept around the world. Schirra displayed none of his early-morning grouchiness as he panned the camera, explaining the various dials and instruments. Towards the end of their second television show, the crew held up a joke sign saying, "Keep those cards and letters coming in folks."

The pressure on the crew to conduct unexpected tests relayed from the ground made Schirra increasingly irate with flight controllers several times, particularly towards the end. He gave them their worst dressing-down in the final 48 hours, on orbit 134, when yet another new test was thrust on the crew. In response Schirra radioed, "I wish you would find out the idiot's name who thought up this test. I want to find out, and I want to talk to him personally when I get back down."

Although he carried out his duties with great skill throughout the flight of Apollo 7, some of Schirra's impatience also rubbed off on Donn Eisele. At one time he became irked over a series of spacecraft control tests that they were to perform, and expressed his

The Apollo 7 crew holds up the cards and letters sign. (Photo: NASA)

annoyance to Mission Control. "I understand the objectives and I understand the reason, but I just don't understand all the changes and so forth at the last minute ... I think it's rather ill-prepared and hastily conceived."

Later on that same orbit, Schirra broke in on Eisele's conversation with the ground, and spoke to Mission Control. "I've had it up here today," he interjected. "From now on, I'm going to be an onboard flight director for these flight updates. We're not going to accept any new games ... or doing some crazy tests we never heard of before. Each test is going to be reviewed thoroughly before we act on it."[16]

As Francis French wrote about Apollo 7 for *In the Shadow of the Moon:*

> Certainly, the vast majority of Schirra's conversations with the ground were polite and professional and he explained any problems clearly ... Schirra did it his way, but he did it well. Despite feeling lousy because of his cold, he rose to the occasion every time the television cameras were switched on, and with his trademark humor gave the viewers quite a show.
>
> Over the decades, the Apollo 7 flight has been unfairly tarnished as one on which the crew 'mutinied,' as if the crew had constantly been hurling insults at the ground. The truth, demonstrated in the transcripts of the space-to-ground transmissions, is very different. They show a crew working carefully and diligently to carry out a thorough test flight of a new machine. There are very few disagreements, and those few were usually accompanied by a clear explanation of why there was an issue. It was not the military precision and exact conformance to orders that had characterized earlier missions. It was the human side of spaceflight: three men working to complete a flight schedule in which they had invested a huge amount of effort. Though each of the crew members sometimes spoke very frankly about something

they disagreed with, it was always about a specific point – usually the addition of a new experiment to the flight schedule.[17]

PRESSING THE ISSUES

Although he did not personally experience any of Schirra's tirades, respected Flight Director Gene Kranz knew that inordinate pressure was being applied to the crew by unscheduled tests. However, he remained unapologetic. "We really put the spacecraft through its paces," he later admitted. "We had a single flight test to do it, and... we kept piling a lot of stuff on Wally, because with only one test to get the job done, every time we saw an opportunity... we'd go for it, we'd press it, we'd try to get some more testing in there."[18]

Further tensions erupted between Mission Control and Schirra when he declared that the crew should go through re-entry with their helmets off. He argued throughout the flight that sinus pressure from their colds (although he was the only one who was really suffering badly from a head cold) could lead to their eardrums bursting; as people with head colds have experienced during steep aircraft descents. He argued the crew should be allowed to pinch their noses and blow to equalize the build-up of pressure – again a very common preventative measure used on descending aircraft. It was a procedure he had requested several times during the mission, but was told that the helmets should be worn for safety reasons. During a final exchange with fellow astronaut Deke Slayton, an irritated Schirra was informed that he would be held to account after the flight for flouting these instructions, but such implications did not concern the testy commander who had already announced his intention to leave NASA and the U.S. Navy following the mission.

The mission voice transmissions went as follows:

Deke Slayton (CapCom): Okay. I think you ought to clearly understand there is absolutely no experience at all with landing without the helmet on.
Schirra: And there's no experience with the helmet either on that one.
CapCom: That one we've got a lot of experience with, yes.
Schirra: If we had an open visor, I might go along with that.
CapCom: Okay. I guess you better be prepared to discuss in some detail when we land why we haven't got them on. I think you're too late now to do much about it.
Schirra: That's affirmative. I don't think anybody down there has worn the helmets as much as we have.
CapCom: Yes.
Schirra: We tried them on this morning.
CapCom: Understand that. The only thing we're concerned about is the landing. We couldn't care less about the re-entry. But it's your neck, and I hope you don't break it.
Schirra: Thank you, babe.
CapCom: Over and out.

Schirra was adamant about not wearing the helmets during re-entry, and he won the argument. Each man took a decongestant tablet about an hour beforehand, and would report no ear problems at all post-flight. On subsequent Apollo flights crewmembers would re-enter without their helmets on.

The astronauts' re-entry began when Schirra triggered eight burns lasting from half a second to 67.6 seconds on the service module's main engine as Apollo 7 made its final sweep over Hawaii during orbit 163. "We're firing right on the mark," he announced, as the craft slowed and began to dip toward the atmosphere. Several minutes later, the service module was jettisoned and the cone-shaped command module orientated itself to face the heat shield on its base to the searing heat as the temperature rose to around 5,000 degrees Fahrenheit as the spacecraft penetrated the ever-thickening atmosphere.

Traveling blunt end forward, Apollo 7 flew across the south-eastern United States toward the planned splashdown zone, 330 miles south-east of Bermuda. Referring to the glowing plasma that engulfed the craft as they swept over the Gulf Coast, Schirra radioed, "We're riding a pink cloud."

Close to the end of its journey, Apollo 7 jettisoned a little cone-shaped "hat" from the top of the spacecraft, exposing its parachute system. Then, at around 24,000 feet two white nylon drogue 'chutes popped out, slowing the craft to nearly 175 miles an hour. Less than a minute later, at 10,000 feet, the three large orange-and-white main parachutes blossomed to slow the heat-seared craft to about 23 miles an hour for the final descent.

Amid sweeping rain showers, the Apollo 7 command module splashed down just a little over a mile from the planned point. On impact the spacecraft turned nose-down in the water, but quickly righted itself as brightly colored flotation bags inflated around the apex. The three fatigued but happy astronauts were soon picked up by helicopter and deposited on the deck of the main recovery ship, USS *Essex* (CV-9) at 8:20 a.m. EST. The command module joined them on deck some 43 minutes later.

Despite being told to "go to hell" during a fiery exchange with Schirra at one point, Flight Director Chris Kraft was full of praise for the Apollo 7 commander. "At times, he

A humorous Benier cartoon published following the splashdown of the Apollo 7 spacecraft. (Illustration: Frank Benier)

At the end of a long and exhausting journey through space, the crew of *Apollo 7* relaxes aboard the recovery ship USS *Essex*. (Photo: NASA)

gave us a hard time during his flight [but] technically what he did was superb. On Mercury, Gemini and Apollo, he flew all three and didn't make a mistake. He was a consummate test pilot. The job he did on all three was superb."[19]

As promised Wally Schirra retired. He was the only astronaut to fly in all three of America's first space programs – Mercury, Gemini and Apollo. Over his three space flights he logged a total of 295 hours and 14 minutes in space.

But as he revealed in a 1998 NASA oral history interview, he was an engineering test pilot who saw no particular enjoyment to be had by spinning endlessly around the planet. He later said of Apollo 7, "I was *bored to tears* up there for 11 days. I mean, bored! Fighter pilots like to fly for an hour, an hour-and-a-half, come back, and do something else. Maybe two flights a day, three flights, then you go to the bar; unless you're going to fly it the next day, then you don't go to the bar. And to sit up there for 11 days, oh, that was so bad! Do you remember those little bands you'd wear around your wristwatch for the calendar? I have that band in a plastic block with 8 of the 11 days scratched off, like a prisoner."[20]

Neither of his Apollo 7 crewmates was destined to fly in space again: Donn Eisele later served as backup Command Module Pilot for Apollo 10 but resigned in 1970. He unexpectedly died of a heart attack at the age of 57 during a 1987 business trip to Japan. Walt Cunningham worked in a management role for the Skylab program with hopes of flying again, but after missing out on a prime crew assignment he resigned from NASA in 1971.

WALLY AND THE MYSTERY BOXES

During the 1960s the Mercury astronauts had enjoyed a special affinity with 1960 Indy 500 winner Jim Rathmann, who ran a Chevrolet and Cadillac dealership in Melbourne, Florida, about 25 miles south of Cape Canaveral. A great admirer of the astronauts, he worked a deal with his friend Ed Cole, the president of Chevrolet, which allowed them to lease brand-new Corvettes for a token $1 per year. They were allowed two vehicles per year and six of them chose classy, speedy Corvettes. The only exception was John Glenn, who preferred a far more sedate and practical Chevrolet station wagon.

When the astronauts returned from their jungle training in Panama in June 1963 they brought with them a baby boa constrictor, knowing that Rathmann hated snakes. When the snake was dropped on his desk at the car dealership he jumped up and bolted out of his office. He returned a few minutes later, saying he appreciated the gag, but he had a pet of his own out back that they might like to see. Wally Schirra takes up the story:

> In the back of his parts shop Rathmann had a wooden box about a foot square by two feet long with heavy gauge chicken wire on one side, a padlocked door, and words of warning stenciled on it: DANGER – LIVE INDIAN MONGOOSE – DO NOT TOUCH. As Shepard watched warily from one side, Grissom and Cooper were beautifully victimized. Rathmann tapped on the box trying to arouse the mongoose, and Gus peered in through the wire. Gordo held the baby boa up to the screen to see

Jim Rathmann won the 1960 Indianapolis 500 and became a friend to the astronauts. (Photo: LAT Photographic)

if the mongoose would react. Of course, the padlock was for show. The door was held by a hook, and when Rathmann flipped it with his finger, a foxtail was released by a spring. Gordo jumped three feet into the air and flung the boa at Gus. To keep it from wrapping itself around his neck, Gus threw the snake into the air. The foxtail landed on a table where Shepard was sitting, and Al started hitting the furry thing with a hammer. Rathmann laughed so hard that he fell against a parts bin, knocking it over and starting a domino reaction of crashing bins.

The mongoose gag got a short reprise in the movie *The Right Stuff*. As Schirra explained, he received credit for the gag because of the movie, but that's incorrect.

"I've pulled it since, but it was Rathmann's idea, and he does it deftly. You must aim the box so the foxtail shoots right through the victim's legs, and you remind him that the mongoose attacks a man at the genitals."[21]

Another story concerning Wally Schirra and a real-life mystery box is told by Elijah Smith from Glasgow, Scotland, who recalls a meeting subsequent to Apollo 7 between his late grandfather Thomas Smith and the astronaut.

The Apollo 1 tragedy was a major setback for NASA and the Apollo program, and in September 1967, North American Aviation merged with Rockwell Standard to form North American Rockwell. Following much grieving, drama and deliberation, the Apollo program was allowed to continue and my grandfather served as the Electrical Supervisor for the construction of the Command Module with North American Rockwell.

After a series of unmanned tests, the first manned Apollo mission, Apollo 7, finally took place between 11 and 22 October 1968. Much was riding on the success of Apollo 7 – the future of the Apollo program and JFK's pledge to reach the Moon before the end of the decade, the future of NASA, the future of North American Rockwell, as well as the Americans' desire to reach the Moon before the Soviets.

Despite a lot of nerves, the mission went as planned and when the astronauts returned they held a meeting with the ground crew from North American Rockwell. Thus officially begins the wee story behind my grandfather's most prized possession.

When in Earth orbit, a spacecraft and its contents experience weightlessness, and when there are bits leftover from the construction of the spacecraft they sometimes emerge from their nooks and crannies and can pose a threat or distraction to the astronauts inside.

During this meeting with the ground crew, Captain Wally Schirra, Commander of Apollo 7, produced a large box containing what he said were the bits found in the Command Module by the astronauts whilst in orbit. My grandfather and his co-workers held their heads in shame. I believe some expletives were uttered when he told the story.

Schirra opened the large box to reveal a slightly smaller box, to the slight relief of the ground crew. And that slightly smaller box contained yet an even smaller box. He proceeded to open the boxes like nesting dolls until the final box, a very small box containing seven small bits of rubbish. In the standards of air and spaceflight, Apollo 7 proved to be an exceptionally clean machine.

The Commander and crew felt the need to express his gratitude to my grandfather by suspending the wee bit of rubbish (some woven insulation from electrical wiring) in a block of translucent plastic mounted to a piece of wood with a brass plaque that reads:

THIS ARTICLE, FLOWN ON
APOLLO 7 OCTOBER 11–22, 1968,
IS PRESENTED TO
T. J. SMITH
BY
CAPT. WALLY M. SCHIRRA
FOR THE CREW OF APOLLO 7

'There are only seven of those in the world,' my grandfather would say.[22]

The plaque presented to Thomas J. Smith by Wally Schirra. (Photo: Elijah Smith)

Wally Schirra was certainly a multi-faceted, larger-than-life personality, and the late former NASA Public Affairs Officer Paul Haney told the author in 2003 he considered himself privileged to have truly known the man over many years.

"Wally was a renowned Mr. Eloquence, and always ready with the incisive final word, but in the forty-odd years I've known him there was only one time when he had nothing to say. That was in the JSC Teague auditorium in 1998, where we had gathered on 1 August for a tribute to Al Shepard, who had died on 21 July. Wally got up on stage, but could hardly say a word [because] he was so choked up, and then he broke down. It was the only time I ever knew this wonderful guy to be lost for words."[23]

REFERENCES

1. Alan Rochford email correspondence with Colin Burgess, 22 July 2015
2. Walter M. Schirra, Jr. with Richard N. Billings, *Schirra's Space*, Quinlan Press, Boston, MA, 1988
3. D. C. Agle, article, "Riding the Titan II," *Air & Space* (Smithsonian) magazine, issue September 1998
4. Gene Kranz, Discovery Channel series *Rocket Science*, Episode 7. Uploaded to YouTube, 6 May 2007
5. Eugene Cernan and Donald A. Davis, *The Last Man on the Moon*, St. Martin's Press, New York, NY, 1999
6. Kathleen Hanser, "Obscure Objects: Tom Stafford's Jingle Bells and Wally Schirra's Harmonica," The National Air & Space Museum, Smithsonian Institution, 16 December 2014
7. Walter M. Schirra, Jr. with Richard N. Billings, *Schirra's Space*, Quinlan Press, Boston, MA, 1988
8. Donald K. Slayton and Michael Cassutt, "*Deke! U.S. Manned Space from Mercury to the Shuttle*," Forge Books, New York, NY, 1994
9. Walter M. Schirra, Jr. with Richard N. Billings, *Schirra's Space*, Quinlan Press, Boston, MA, 1988
10. *Ibid*
11. Clayton Moore, *Airport Journals* article, "Levity Beats Gravity: Remembering Pilot and Pioneering Astronaut Wally Schirra," online at: *http://airportjournals.com/levity-beats-gravity-remembering-pilot-and-pioneering-astronaut-wally-schirra*
12. Walter M. Schirra, Jr. with Richard N. Billings, *Schirra's Space*, Quinlan Press, Boston, MA, 1988
13. *Ibid*
14. Walter M. Schirra, Jr., interviewed by Roy Neal for NASA JSC Oral History program, San Diego, CA, 1 December 1998
15. Francis French and Colin Burgess, *In the Shadow of the Moon*, University of Nebraska Press, Lincoln, NE, 2007
16. Sydney *Daily Mirror* newspaper, "Those Angry Young Men in Apollo 7," issue 1 November 1965
17. Francis French and Colin Burgess, *In the Shadow of the Moon*, University of Nebraska Press, Lincoln, NE, 2007
18. *Ibid*
19. Clayton Moore, *Airport Journals* article, "Levity Beats Gravity: Remembering Pilot and Pioneering Astronaut Wally Schirra," online at: *http://airportjournals.com/levity-beats-gravity-remembering-pilot-and-pioneering-astronaut-wally-schirra*
20. Walter M. Schirra, Jr., interviewed by Roy Neal for NASA JSC Oral History program, San Diego, CA, 1 December 1998
21. Walter M. Schirra, Jr. with Richard N. Billings, *Schirra's Space*, Quinlan Press, Boston, MA, 1988
22. Elijah Smith email correspondence with Colin Burgess, 12–13 June 2015
23. Paul Haney email correspondence with Colin Burgess, 21 September 2003

Epilogue: Final salute to a hero

Prior to leaving NASA, Wally Schirra narrated a 12-inch album about the Apollo 11 mission called *Flight to the Moon*, which was released after his retirement. The first side was called *Moments in Space History* and detailed all of NASA's manned space missions up to the first lunar landing, while the reverse side had as its title *The Actual Voyage of Apollo 11*. "That's what gave me enough money after taxes to buy that Maserati downstairs," he told one reporter with a chuckle in his voice.

In 1967, he was offered a position as a director with Imperial American (Oil and Gas) by oil tycoon John M. King, who owned more airplanes than most small airplane companies. A flamboyant Colorado millionaire entrepreneur, King was a large man given to wearing cowboy hats and boots, who was once described as having built his empire on the sound American business principle of tax avoidance through oil leasing and other perfectly legal means.

"I gave King a ride once in a T-38," Schirra recalled with a smile. "Not legal, of course, but this sort of thing is now past the statute of limitations. We buzzed King's ranch in Colorado, and blew the windows out!"

Although Schirra worked for King in Colorado, he did not fly any of the airplanes in the fleet. "At first I tried to fly," he said. "King had enough airplanes to keep me busy – a Jetstar, Electra, Lears, twin Navions with rocket assist – but flying and talking to clients didn't mix. When you fly, you can't carry on a business deal. I had to make a choice, and left the flying to the corporate pilots."

"I did own a Hughes 300 helicopter for a while," he added. "In the space program we all had to have time in a helicopter, simulating the stage when we orbited the Moon, and had to transition from horizontal flight to vertical flight." He flew the helicopter at the time he was working for John King, but at 5,000 feet and up, it was barely able to handle the thinner air. "I was pushing my luck and I finally got rid of it."[1]

Afterwards Schirra bought a turbocharged Enstrom that was far better suited to the high Colorado Rockies. But when he and Jo moved to Rancho Santa Fe in Southern California in 1984 and enquired about landing the helicopter near his house, he found his new neighbors far from agreeable. The helicopter had to go, and it did.

C. Burgess, *Sigma 7*, Springer Praxis Books, DOI 10.1007/978-3-319-27983-1

He later said he regretted his association with John King, whose empire began to crumble shortly after Schirra left the company. "I learned all the do-nots from that outfit," he said.

In his post-NASA years, Wally Schirra became a commentator for CBS News and covered a number of significant space flights, often teamed up with veteran anchorman Walter Cronkite. Together they covered all six Apollo landings on the Moon, plus the dramatic abort of Apollo 13. Their biggest moment together though, was undoubtedly mankind's first footsteps on the Moon during the Apollo 11 mission. That caused both men to become quite emotional on live television.

> Oh, that was tough, yeah, I got caught, in fact... They caught me and actually a tear came into my eye. I didn't even know it happened, it was that emotional.

Although no caption was supplied with this photo, it seems as if Schirra is celebrating the Apollo 11 lunar landing in 1969. (Photo: Retro Space Images)

Walter Cronkite and Wally Schirra commentating for CBS on the successful landing of Apollo 11 on the Moon. (Photo: CBS)

About three days before they landed, I said, "Walter, there's no man I know of, other than possibly your contemporary Eric Sevareid" – I was needling him about Eric, because Eric is *so* well spoken – "who could possibly think of something brilliant to say when they land on the Moon. Now here we are, we've been waiting for eons literally to land on the Moon. We know it's going to happen, the guys have a good mission, things are going on very well. What are you going to say when these men land on the Moon, when they step foot on the Moon, what are you going to say, Walter? I can't think of it, I'm tired of saying fantastic, beautiful, all this kind of stuff." Walter said, "I don't know, I think I'll address myself to that." So anyway, when the tear's coming down out of my eye, Walter's saying '*Golly gee whiz*!' Famous quote. Which just shows how emotional it was.[2]

In 1969 Schirra also ventured into private industry as president of Denver-based financial company Regency Investors, Inc., and served from 1970 to 1972 as chairman and CEO of the Environmental Control Company. In January 1973 he was elected vice chairman of the board for SERNCO, Inc., then promoted to chairman in July. He was slightly burned in an association with a questionable investment company to which he lent his name, but treated it as a lesson learned.

He was also invited to become a director of Investors Overseas Services (Australia) Management Limited. But he later said that even though he loved Australia, "I learned a little about how Australians did business [because] I was in the company of IOS fund salesmen for the length of my visit. They were a brassy bunch, ostentatious and rude. They wore suede shoes and big diamond rings and rode around in Rolls-Royces. I left Australia on the first available flight."[3]

LIFE BEYOND NASA

Schirra once told an interviewer interested in the way he had gone from astronaut to the world of business that he had "no formal business training whatsoever."

> I was a professional naval officer. But then I became an astronaut and when you put your life at stake and depend on industry for support, you get to know what it is all about very quickly. I soon learnt how the industry worked.
>
> Of course, being an astronaut helped immensely. My mind is not confined to a tight formula and I have had plenty of experience of making decisions on a high level. I do not worry about the petty details. I leave those to the men out of Harvard Business School. You need "bean counters" as I call them, to keep track of all the decimals. I make decisions on a broader scale.[4]

There were other business and charitable interests: he became vice president of the Society of Experimental Test Pilots; chairman of the Colorado State Cancer Crusade; a director of Precision Estimates – a data storage company for computers; a director of the First National Bank of Colorado; a trustee of the Detroit Institute of Technology; a spokesman for the Association of American Railroads; and a director of the U.S. Winter Athletics Committee.

In 1975 he was appointed director of technology purchase for the Johns Manville Corporation, and the next year became their Director of Power Plant and Aerospace Systems. Three years later he became president for development with the Goodwin Corporation. The year after that he established his own company, Schirra Enterprises, as an independent consultancy.

In 1981, on the occasion of the first space shuttle flight, Schirra was asked for his thoughts on the innovative spacecraft. "Mostly it's lousy out there," he stated. "It's a hostile environment, and it's trying to kill you. The outside temperature goes from a minus 450 degrees to a plus 300 degrees. You sit in a flying Thermos bottle."[5]

Three years on, in 1984, the Schirras moved to Ranch Santa Fe, outside of San Diego, California. They were both familiar with (and fond of) the area. During Wally's two assignments as a test pilot at Miramar Naval Air Station in the 1950s they had lived in Solana Beach and Del Mar, while Jo had lived in Coronado as a child with her military family.

Living a contented life with Jo in Rancho Santa Fe, Wally loved nothing better than sailing on the *Windchime*, his 36-foot sailboat. He also enjoyed skiing, hunting, and fishing, and stayed active as a popular orator on the guest speaker circuit. He and Jo were frequent participants in social and charitable activities. Wally also served as a member of the San Diego Yacht Club and the Rancho Santa Fe Tennis Club and was on the boards or advisory boards of several organizations, including the Birch Aquarium, the Salk Institute, and Sharp Hospital. One of his greatest prides in his later years was serving as a board member of the San Diego Air & Space Museum, whose space flight gallery is named after him.

Another great source of pride for Wally Schirra was the Astronaut Scholarship Foundation. The surviving Mercury astronauts decided to use their joint credibility to encourage students to pursue scientific endeavors in order to help to keep America on the

Jo and Wally Schirra were VIP guests at the launch of Space Shuttle *Columbia* from the Kennedy Space Center on 4 April 1997 for mission STS-83. (Photo: NASA KSC)

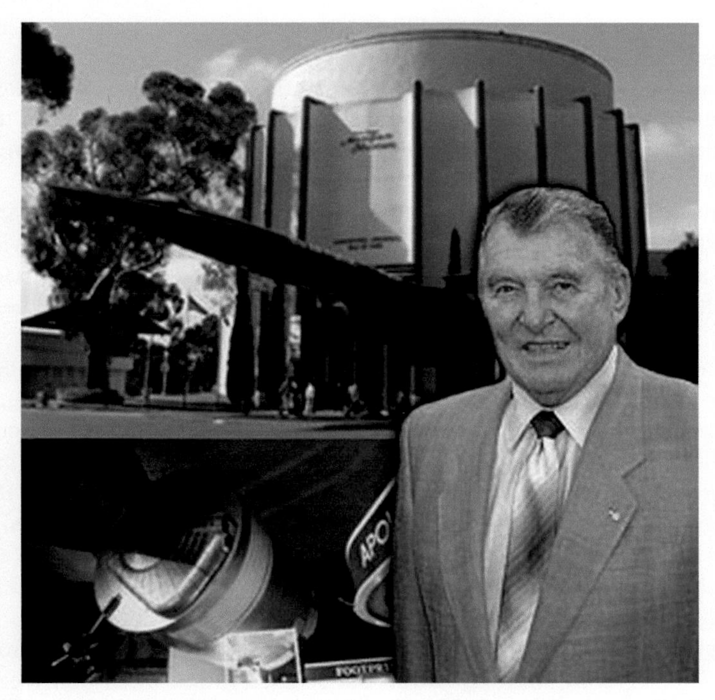

The San Diego Air & Space Museum in Balboa Park. Schirra had a great affection for this museum, and used to jokingly refer to it as "Wally's World." (Photos: San Diego Air & Space Museum)

leading edge of technology. That idea led to the formation of the Mercury Seven Foundation, later to become the Astronaut Scholarship Foundation. Along with Betty Grissom (widow of Gus Grissom), William Douglas, M.D. (the Project Mercury flight surgeon), and Henri Landwirth (Orlando businessman and friend of the astronauts) the six Mercury astronauts provided scholarships for students who excelled in the area of science. For a 2004 interview Schirra commented:

> We [formed] what was called the Mercury Foundation, then we made it the Mercury Scholarship Foundation. Now we call it the Astronaut Scholarship Foundation. And with giving speeches and other charitable events, we now have – this is unreal – 20 students a year at $10,000 each. $200,000 a year out of our foundation for scholarships. We pick them up at junior level in college or university. We have nothing to do with picking the people – we just assign people to do that. We arrange or provide the money. And we have over $2 million in our bank account for this project, so we're working with a company called Delaware North that runs the Kennedy Space Center Visitor Center and with enough attendance at our Astronaut Hall of Fame, we have funded now [a] little more than $200,000 a year in support of our scholarship fund. That's the best plaque I'll ever have. We have had over 42 go on to doctorate. Isn't that a good feeling?[6]

A 25th anniversary photo acknowledging the work of the Mercury Seven Foundation, now the Astronaut Memorial Foundation. At back, front left: Gordon Cooper, Scott Carpenter, Alan Shepard, Wally Schirra and Deke Slayton. At front: John Glenn and Betty Grissom, widow of Gus Grissom. (Photo: Astronaut Memorial Foundation)

Wally Schirra was enshrined in the National Aviation Hall of Fame in 1986 and was inducted into the U.S. Astronaut Hall of Fame in 1990. In 2000 he was also inducted into the Naval Aviation Hall of Honor. Five years after that, in 2005, he was named a NASA Ambassador of Exploration.

He was often asked about writing a book on his life's experiences, and in 1988 the autobiographical account *Schirra's Space* (co-authored with Richard N. Billings) was released. In 1995 he also shared authorship with Richard Cormier, Phillip Wood and Barratt Tillman in the Phalanx book *Wildcats to Tomcats: The Tailhook Navy*, and in 2005 he co-authored *The REAL Space Cowboys* together with his long-time friend Ed Buckbee.

Wally Schirra was always content and proud of what he had accomplished as a NASA astronaut, but had no desire to return to space as a 'passenger' in the way John Glenn flew a space shuttle mission in 1988. Although, typical Schirra, he did qualify that remark by pointing out that he would only go if he could pilot the vehicle. "I was amazed, frankly, when John Glenn, who had only five hours in space, was anxious to go up there for – what was it? – 8.3 days or 9.3 days," he stated. "I was bored to tears up there for eleven days. I mean, bored! Fighter pilots like to fly for an hour, an hour-and-a-half, come back, and do something else. Maybe two flights a day, three flights, then you go to the bar; unless you're going to fly it the next day, then you don't go to the bar. And to sit up there for eleven days, oh that was so bad!"[7]

In a media interview given just a month before his death, the ecologically passionate Schirra repeated a phrase that he had often used in his speeches over many years: "I left Earth three times. I found no place else to go. Please take care of Spaceship Earth."[8]

The ebullient, much loved and ever-wisecracking Wally died on 3 May 2007. His widow Jo later confirmed to me that he had a heart attack while undergoing treatment for abdominal cancer at the Scripps Green Hospital in La Jolla, California. He was 84 years old. His sister Georgia Lou (Burhans) had predeceased him in April, 2000, aged just 73.

Wally Schirra was enshrined in the National Aviation Hall of Fame in 1986. His accomplishments also earned him induction into the U.S. Astronaut Hall of Fame in 1990, and the Naval Aviation Hall of Honor in 2000. (Photos: Wally Schirra personal collection, San Diego Air & Space Museum)

Ed Buckbee and Wally Schirra, co-authors of the book *The REAL Space Cowboys*. (Photo: www.wallyschirra.com)

Mercury, Gemini and Apollo astronaut, Walter M. Schirra Jr. (Photo: NASA)

Capt. Wally Schirra's ashes, along with those of eight other U.S. Navy veterans were committed to the sea on 11 February 2008 during a special burial ceremony aboard the USS *Ronald Reagan* (CVN-76).

The ashes of Wally Schirra (the ones being held) and those of eight other Navy veterans are respectfully committed to the sea. (Photo: U.S. Navy)

In speaking of his Mercury colleague and friend, Scott Carpenter took time to reflect on a superb pilot and inveterate prankster. "He was a practical joker, but he was a fine fellow and a fine aviator. He will be sorely missed in our group."

"I thought he was the best of the first seven astronauts," said fellow astronaut Bill Anders of the Apollo 8 crew. "He was easy to work with, competent and he made no errors. The only shortcoming he had was that he told dumb jokes."[9]

"Wally was the most friendly and outgoing of us all," said Walt Cunningham, who flew the Apollo 7 mission with him. "But when it was necessary, he was completely serious. He could put his foot down, and he was an extremely fine aviator."[10]

Even after his death, Schirra would be awarded honors. On 20 October 2008 he was posthumously awarded NASA's Distinguished Service Medal for his Apollo 7 mission. In San Diego the following year, on 9 March 2009, a crowd estimated at around 1,500 applauded with enthusiasm as Jo Schirra broke a traditional bottle of champagne over the bow of the 689-foot-long dry cargo/ammunition ship which was launched into San Diego Bay as the USNS *Wally Schirra* (T-AKE 8) in honor of her late husband. The senior Navy astronaut on active duty, Captain Lee Morin, M.D., Ph.D., served as the ceremony's principal speaker, and General Dynamics Chairman and Chief Executive Officer Nicholas Chabraja also spoke. Also present were ex-astronauts Bill Anders, Scott Carpenter, Jim Lovell and Tom Stafford. The U.S. Military Sealift Command took delivery of the new ship on 1 September that year.

Jo Schirra breaks the customary bottle of champagne over the bow as the USNS *Wally Schirra* (T-AKE 8) prepares to slide into San Diego Bay. (Photo: Shalene Baxter)

The Schirra family wave as the mighty ship is officially launched. (Photo: Shalene Baxter)

Marty, Jo and Suzy Schirra after the USNS *Wally Schirra* has been launched. The ship can be seen turning in the background (Photo: Francis French)

Dry cargo/ammunition ship USNS *Wally Schirra* at sea for the first time in August 2009 conducting sea trials off the coast of San Diego. (Photo: General Dynamics NASSCO)

The author in October 2015 alongside the plaque on the Mount Soledad memorial wall honoring the late Capt. Wally Schirra, USN. (Photo: Jennifer Brisco)

With family members in attendance, the late naval aviator/astronaut was honored by a memorial plaque on 7 November 2009 on Soledad Mountain, overlooking San Diego. During the Mount Soledad Memorial Association's Veteran's Day ceremony, a black granite plaque was unveiled on one of the memorial walls. Radio talk show host Mark Larsen, a personal friend of Schirra, led the moving ceremony. The commander of the Navy Region Southwest, Rear Adm. William French, delivered an address. The Navy Band Southwest then paid a musical tribute and the San Diego Salute T-34 Formation Team performed a flyover.

On 27 April 2015, Jo Schirra, Wally's beloved wife of 62 years, passed away peacefully in her Rancho Santa Fe, California home with her family by her side. She was 91 years old. At Jo's request, her ashes were sprinkled into the Pacific Ocean off San Diego and Kauai. She is survived by their two children, Walter III (Marty) and Suzanne.

A WELL-TRAVELED SPACECRAFT

McDonnell's Mercury Spacecraft No. 16, named *Sigma 7*, was returned to Hangar S at Cape Canaveral for post-flight work and inspection. It was then planned to retain the spacecraft at Cape Canaveral for permanent display. First of all, it was returned to the McDonnell plant in St. Louis, Missouri for a further inspection by the company's own engineers and technicians, after which it was placed on display there for a short time.

The *Sigma 7* spacecraft loaded into a truck for the journey to the McDonnell plant in St. Louis. (Photo: NASA)

Sigma 7 then went on a limited world tour, which included spending display time in South America. The previous year, late 1962, NASA had received a request for support of the 1st International Space Exposition to be held at a major recreation and park site, Ibirapuera Park, in Sâo Paulo, Brazil. Great Britain, France and the Soviet Union were also to participate in the show.

As the person charged with organizing the NASA exhibits, Ellwood Johnson would later say, "NASA wanted to be a major player in the Exposition. They secured Wally Schirra's Mercury spacecraft, a full-scale model of the X-15 research rocket plane from the U.S. Air Force, and an array of exhibits that were developed for the exposition. In late January 1963, I left for Sâo Paulo, Brazil … We had set up a month's training program. A lot of time was spent trying to figure out what the Russians were going to do; when push got to shove, they decided not to participate. Finally, a U.S. Air Force cargo plane arrived with Wally Schirra's *Sigma 7* spacecraft, a full-scale model of the X-15 research plane (50 feet long and 22 feet wide) and the exhibits. The X-15 was accompanied by an Air Force exhibit team who remained with the model throughout the exhibition. We set up everything at the exposition and were ready for the opening. Incidentally, the Brazilians were sure that the *Sigma 7* was a model and that the X-15 was for real."[11] The spacecraft would be on temporary display at the Exposition from 15–26 March 1963. It was later shipped to Paris for a science exhibition in the Palais de la Découverte (Discovery Palace) during June 1963.

Sigma 7 on display at the McDonnell plant. (Photos: NASA)

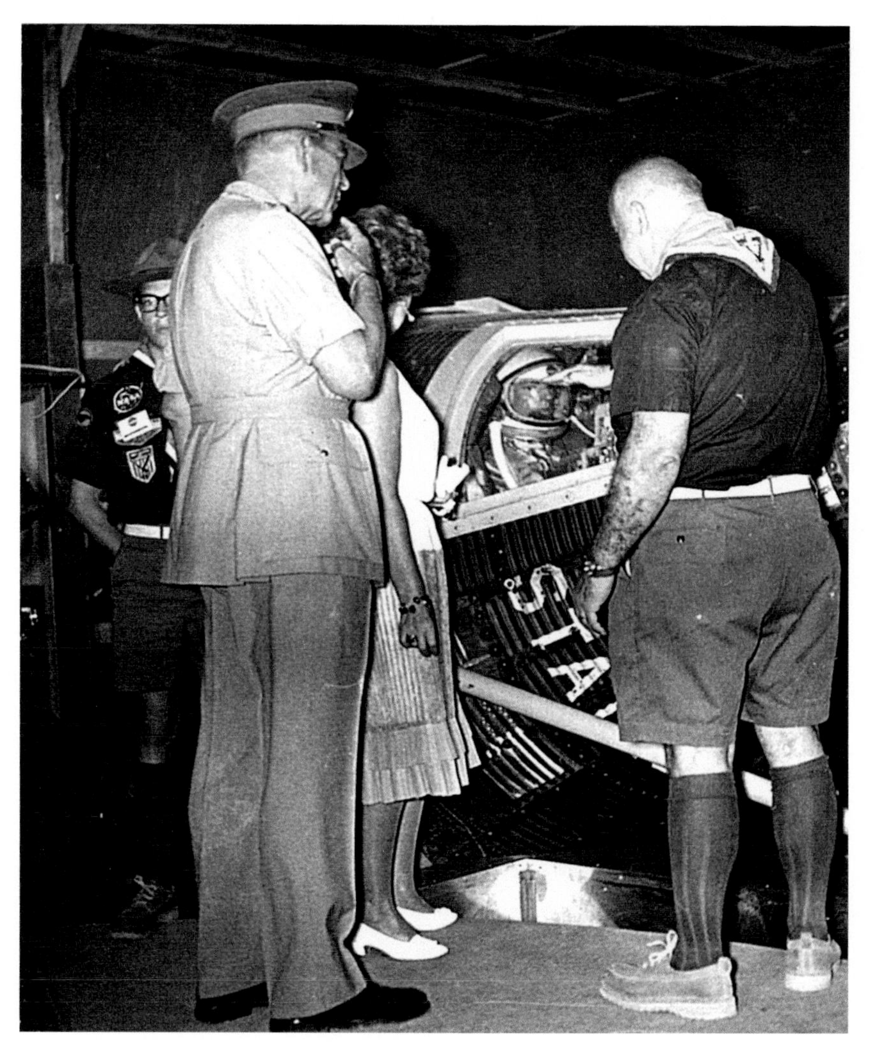

The spacecraft was a major attraction at the World Scout Jamboree in 1963. In this photo King Paul and Queen Federica are shown *Sigma 7*. (Photo: Konstantinos Dassios, marathon1963.com)

From August 1–11 of that same year, an interesting display of the spacecraft took place at the 11th World Scout Jamboree at the Plains of Marathon, outside of Athens, Greece. NASA had cooperated with the Greek Armed Forces to exhibit *Sigma 7* at the Marathon Jamboree. A dedicated hangar was constructed, and there were lengthy lines to view the spacecraft. In October 1963 the spacecraft traveled to Orlando, Florida, as part of a display on Space Science Achievements.

Sigma 7 arrives at the Marathon exhibition site (top) and scouts and visitors waiting to view it (bottom). (Photo: Konstantinos Dassios, marathon1963.com)

Wally Schirra receives a scale model of *Sigma 7* from James S. McDonnell. (Photo: NASA)

On 15 November 1963, Wally Schirra visited the McDonnell plant in St. Louis and was presented with a scale model of *Sigma 7*, complete with Cece Bibby's logo on its side.

That same month, *Sigma 7* was sent on another world tour. Among other European venues, from 7–24 November it was on display at the RAI Exhibition and Convention Centre in the Dutch capital of Amsterdam.

While on exhibition at the U.S. Space & Rocket Center in Huntsville, Alabama, people were able to touch the historic spacecraft. (Photo: NASA Marshall Space Flight Center)

The well-traveled spacecraft would continue to be displayed at various venues at home and aboard. For example, from 10–20 March 1966 it featured at the National Orange Show in San Bernardino, California as part of an indoor display in the Swing Auditorium provided by NASA. And in May it was on show at the Peaceful Uses of Space Conference in Boston, Massachusetts.

On 20 September 1967 NASA transferred title of *Sigma* 7 to the Smithsonian Institution while the spacecraft was on exhibit at the NASA Marshall Space Flight Center in Huntsville, Alabama. It remained on exhibit there for the next 27 years.

In March 1990 the spacecraft was moved to the U.S. Astronaut Hall of Fame in Titusville, Florida, where Schirra is honored as one of the Hall's original inductees. Located not far from the Kennedy Space Center, the Astronaut Hall of Fame was an offshoot of the U.S. Space & Rocket Center in Huntsville. Some years ago the Hall was on the verge of failing and was taken over by the KSC Visitor's Center. At the time of writing, the Hall is due to close and the contents, including *Sigma 7*, will be relocated to the main Visitor Center.[12]

Whilst the immediate future of the historic *Sigma 7* spacecraft is known, mystery still surrounds it, as indeed it does Scott Carpenter's *Aurora 7*, because the shingled panel on which artist Cece Bibby painted the spacecraft's logo appears to have been wholly replaced by another, blank shingle. Whether this is the result of the artwork shingles having been removed long ago for preservation purposes and subsequently misplaced, or through theft, is sadly still not known at the time of writing this book.

During a visit to the U.S. Space & Rocket Center in Huntsville, Alabama in 1983, the irrepressible Wally Schirra decided to have one more ride on his *Sigma 7* spacecraft. Apparently he tried to find a cowboy hat for the photo but was unsuccessful. (Photo: NASA Marshall Space Flight Center)

Sigma 7 on display at the U.S. Astronaut Hall of Fame in Titusville, Florida. (Photo: Wikipedia)

REFERENCES

1. Freia I. Hooper, article for unknown newspaper from Wally Schirra personal papers, San Diego Air & Space Museum, issue October 1990, pg. 111
2. Jerry Bledsoe article, "Down From Glory," *Esquire* magazine, issue January 1972, pg. 182
3. Walter M. Schirra, Jr. with Richard N. Billings, *Schirra's Space*, Quinlan Press, Boston, MA, 1988
4. Michael Morton-Evans, Sydney *Daily Mirror*, "The spaceman who found a place on Earth," issue 22 October 1969
5. Thomas Watkins, Associated Press obituary article, "Astronaut Schirra laid path for Apollo," 4 May 2007
6. Fran Foley interview with Wally Schirra for Veterans History Project, 19 April 2004
7. Wally Schirra interview with Roy Neal for NASA JSC Oral History program, San Diego, CA, 1 December 1998
8. Associated Press interview with Wally Schirra, April 2007
9. Michael Kinsman, *Union-Tribute* (San Diego) newspaper article, "Most human of all astronauts," issue 4 May 2007
10. *Ibid*
11. Ellwood A. Johnson, extracts from his unpublished article, The NASA Years, 1962–1965, online at: *http://www.lowrymn.com/ellwood%20a.%20johnson.htm*
12. National Air and Space Museum (NASM) Smithsonian Institution, *McDonnell Mercury Spacecraft #16 Accession Worksheets, 20 September 1967 – 19 March 1990*, courtesy of Michael Neufeld, Senior Curator, NASM Space History Division

Appendix 1

Comparative countdown summary from astronaut waking

	MA-6 (Glenn)	MA-7 (Carpenter)	MA-8 (Schirra)
Pilot awaked	2:20 a.m.	1:15 a.m.	1:40 a.m.
Nominal launch	8:00 a.m.	7:00 a.m.	7:00 a.m.
Time to nominal launch	5 hrs 40 mins	5 hrs 45 mins	5 hrs 20 mins
Actual launch	9:47 a.m.	7:45 a.m.	7:15 a.m.
Time to actual launch	7 hrs 27 mins	6 hrs 30 mins	5 hrs 35 mins

Sequence of events

Event	Pre-flight predicted time hr:min:sec	Actual time hr:min:sec
Launch phase		
Booster engine cut-off (BECO)	02:10.8	02:08.6
Tower release	02:33.8	0:02:33
Escape rocket ignition	02:33.8	0:02:33
Sustainer engine cut-off (SECO)	–	05:15.7
Tail-off complete	05:05.8	05:15.9
Orbital phase		
Spacecraft separation	05:06.8	05:17.9
Retrofire sequence initiation	50:21.8	8:51:30
Retrorocket (left) No.1	50:51.8	8:52:00
Retrorocket (bottom) No.2	50:56.8	8:52:05
Retrorocket (right) No.3	51:01.8	8:52:10
Retrorocket assembly jettison	51:51.8	8:53:00

C. Burgess, *Sigma 7*, Springer Praxis Books, DOI 10.1007/978-3-319-27983-1

Re-entry phase

0.05g relay	00:20.8	9:01:40
Drogue parachute deployment	05:36.8	9:06:50
Main parachute deployment	07:02.8	9:08:12
Main parachute jettison (in water)	11:33.8	9:13:11

Appendix 2

NASA Report: Changes incorporated into *Sigma 7* spacecraft (Reproduced from NASA MSC *Space News Roundup*, issue 17 October 1962)

According to a NASA report in the 17 October 1962 issue of the Manned Spacecraft Center's *Space News Roundup*, a number of significant changes were incorporated into the *Sigma 7* spacecraft for the MA-8 mission, as increased experience with manned space flight continued to improve systems and hardware used on successive flights.

Greater flexibility in the manner in which the pilot could control the spacecraft's attitude; modifications in the communications capabilities in and out of the spacecraft; and an increase in the number of temperature sensing gauges were among the changes.

A panel switch permitted Schirra to set controls to give only low thrusters operation during fly-by-wire mode, thus conserving fuel.

In press conferences following the flight, Schirra used the term "fly-by-wire low." Explaining, he said, "This only means that we have changed since John's and Scott's flights. We have added a switch to permit us to use the very low thrusters for maneuvering while in orbit. This made it very much easier for me to handle. In Scott's case, he had a little more trouble with the high thrusters cutting in frequently. This way, we could cut them out and I could concentrate completely on control and continue with other tasks simultaneously."

In order to provide the astronaut with communications if he left the spacecraft and entered his life raft after landing, as Scott Carpenter did, Schirra carried a small but high-powered radio transceiver which in tests successfully made contact with an aircraft flying at 9,000 feet about 19 miles away.

A special extension cable was also added to the much higher powered radio transmitter aboard the spacecraft. By throwing the extension cable out through the upper neck of the spacecraft, Schirra could have used the spacecraft radio while sitting in the life raft. The cable idea is aimed at providing communications in the event of emergency or contingency landing many miles from pre-planned landing areas.

C. Burgess, *Sigma 7*, Springer Praxis Books, DOI 10.1007/978-3-319-27983-1

Space-age "rabbit ears" have been added to the lower section of the spacecraft. A new high frequency dipole antenna which is coiled to the diameter of two silver dollars but extends 14 feet on each side of the retrorocket package in flight gave "tremendous increase in reception," Schirra said, so much so that on one occasion he had perfect communications with Quito, the Minitrack station in Ecuador.

"We had much better reception as far as transmissions ranges. Acquisition times for HF transmissions from *Sigma 7* must be reduced later from post-flight data. I felt on occasion that I was raising stations earlier than I had anticipated. But I'm convinced we really are making progress on communications with this antenna."

Temperature sensing instruments have been attached to the center of the metal dome on the heat exchangers of both pressure suit and cabin environmental control system. A direct reading temperature gauge gives the pilot a continuous reading on both cooling devices during flight. Schirra commented that the device gave "a very accurate readout on the suit temperature circuit."

Other temperature pick-ups were installed on the cabin heat exchanger outlet and the inside of the suit circuit where oxygen flows into the suit.

The addition of a temperature-survey indicator and a 12-position rotary switch to let the pilot monitor heat exchanger air outlet temperatures, 3-inverter temperatures, right retrorocket temperatures and thruster temperatures completes the new system.

In other changes, leg support troughs have been eliminated from the couch inside the spacecraft cockpit. A lateral knee support consisting of plates outside of each knee, to keep the knees from spreading outward during acceleration, has been substituted. The plates can be locked into place or retracted at the pilot's discretion.

Commented Schirra, "This gave me quite a bit more room to move my legs around. In fact, I could sit in there while waiting during the countdown period with my legs crossed."

The small toe cups of earlier spacecraft have been replaced by a larger cup, and the heel support built up to provide restraint for the lower legs and feet.

These changes were designed to afford greater mobility, improved circulation and permit easier access and egress from the spacecraft.

The magnetic tape used in the spacecraft has been changed to a thin-base type which increases the possible recording time from six to eleven hours.

Schirra also spoke of increased readouts on the electrical system. He mentioned an "attitude select" switch that allowed him to select either re-entry or retro attitude during flight, "which meant I was flying at zero pitch."

Appendix 3

Francis French is currently the Director of Education at the San Diego Air & Space Museum in southern California. One person who was influential in this appointment was the subject of this book, Wally Schirra, who was on the board of the museum.

Francis and I have known each other as great friends for some three decades now, and over those years we have each observed the other expand that interest to where it stands today. Several years ago, after I had been appointed as the series editor for the University of Nebraska Press's *Outward Odyssey* books on the history of human space exploration, Francis and I co-authored the first two volumes in that series: *Into That Silent Sea* and *In the Shadow of the Moon*. For the first volume, Francis was charged with writing the chapter on the life and accomplishments of Wally Schirra, and he was able to utilize an extensive interview he had conducted with the former astronaut in February 2002.

While the interview certainly covers a lot of ground already explored in this book, it was felt that Wally's unique sense of humor, and his incisive opinions on where things stood back then and where he thought we might be heading in the future are well worth repeating in this book.

I worked with NASA, not for NASA

An interview with astronaut Walter "Wally" Schirra
by Francis French, 22 February 2002*

It has been said that if you go out to lunch anywhere in the nicer neighborhoods around San Diego, California, you'll probably be within listening distance of a former Navy Captain. If you are lucky, one of the people you'll see telling old tales in the corner with a mischievous grin will be Captain Wally Schirra, one of the original Mercury Seven astronauts. Schirra is the only person to have flown a Mercury, Gemini and Apollo mission, and thus can present a unique view of the earliest days of manned spaceflight – tales which are now slipping from common memory into the realm of written history. Recently, my wife

C. Burgess, *Sigma 7*, Springer Praxis Books, DOI 10.1007/978-3-319-27983-1

and I had the opportunity to sit back with Schirra in his favorite lunchtime spot, and discuss some of those stories first-hand.

Wally Schirra has been portrayed by many people who remember those early missions as a prankster, a practical joker and a pun maker of renown. Decades after the stories began, it is delightful to see he is still at it. Neither we nor any waitresses close by escaped the impish plays on words, delivered with a hearty chuckle. He rarely missed the chance to light-heartedly rib his fellow astronauts about incidents they'd probably rather forget.

Like others in the Mercury 7, Schirra started flying at a very early age. His father was a World War One flying ace, and his mother had been a wing-walker, so it was a good family for a future pilot to grow up in. Being around airplanes also meant he got to meet many of the famous early aviation pioneers.

"Dad went to Canada, to learn how to fly with the Royal Canadian Air Force. He was commissioned a First Lieutenant, not a Second Lieutenant, when he finished. He often ribbed me in later years that it took me three years to become the same rank that he was when he was commissioned! He came to Texas, instructed there, and then went and flew with the Royal Flying Corps, the precursor to the RAF, before the United States joined World War One. So he had quite an interesting background. He took me on my first airplane ride, where I could have a hand on the stick. I recall going down to Teterboro, New Jersey and aviation pioneer Clarence Chamberlin took me up on a monster airplane. I also remember seeing war hero Jimmy Doolittle fly a Gee Bee racer there. He was my childhood hero. Many years later, in 1969, I was lucky enough to go hunting with him, in Wyoming, for a week."

Initially, Schirra was not too interested in a military career – as a late teen, he was more interested in tinkering with sports cars, dating women, and trying to play swing music on the trumpet. The Japanese bombing of Pearl Harbor made him change his mind, and he decided to take the test to go to West Point and begin a career in the Army. However, when sitting down to take the test, he discovered that Army candidates had to take an additional history test, so he quickly changed his application to one for the Naval Academy instead.

"I changed all the headings, yeah! I changed my mind at the last minute. But, truth be told, I had always wanted to go to the Navy. As a young kid, I was intrigued by a Naval Officer with the beautiful brown shoes and sharp gold wings walking through my home town, so that too had made an impact."

Schirra graduated from the Naval Academy at Annapolis, and while visiting his parents he visited the Army-Navy Country Club. It was there that he met his future wife, Jo. Being wartime, they had only seven days to get to know each other before Schirra was assigned to a ship in Hawaii.

"Before I left for the Pacific, I picked my girlfriend there at the old Navy Country Club, and then suddenly I'm gone to a ship. She had only dated P-51 pilots, but I was a black-shoe, I don't think I'd even flown anything yet! Her father was a Navy Captain at the time. It was a classic wartime romance. We corresponded almost every day by mail. I came back and said, we're getting married – cancel all those other dates! It worked! And it's worked for quite a while."

Once the Second World War was over, the newly married Schirra became the first of his academy class to enter flight training. Having become a naval aviator, he made ninety combat flights in Korea, shooting down two MiGs. He was hoping to be assigned to the test pilot school, but was instead assigned to China Lake, to test the Sidewinder air-to-air missile.

"That was the early days, in the California desert after Korea, my first tour in California in fact. An interesting time for me, we developed a weapon that would chase heat – that's why they're called a heatseeker. I saw it first as a mockup. I had a cigarette in my hand, and this little gold thing, an optical device, is following me around the room. I thought, whoops, what is this, knowing a jet engine is a lot hotter than a cigarette. And that was the beginning of Sidewinder. Interesting weapon, still in production, though modified now.

"One time, while testing the Sidewinder, I fired one, and it started doing a loop. I said, 'I'm not sure where it's going, but I'm not going to lose sight of it.' So I was flying inside, I had to do a loop inside of its loop – and it burned out. You're not sure until it's dead. It might burn out before it got to me, but I wasn't about to wait for that to happen!"

Eventually, Schirra got his wish, and was assigned to the Navy's test pilot school at Patuxent River, Maryland. He studied aerodynamic theory and worked closely with engineers, experiences that no doubt contributed to what happened next in his career. He was given orders to report to the Pentagon, with no reason given. On arrival, he discovered that he was being considered as a candidate for Project Mercury. Schirra was very skeptical at first, seeing it as a career interruption that he could not afford.

"As a child I went to a circus in New York City. They had a man shot out of a cannon into a net. At this briefing I said, that's who you want. You don't want one of these engineering test pilots. You just want a dummy to get into a capsule on top of a rocket – I'm not going to do that! In contrast to Glenn, who said, 'I want to go now!' Then, as I talked to my peer group, they kept saying, 'Do you want to go higher, farther and faster? This is the way to go about it.' We talked about how space was a totally different environment. So I went along with it for a while, and finally became intrigued with what was going on, and realized what it was."

In his book *The Right Stuff,* Tom Wolfe speculated that the Mercury 7 did not turn the project down as it was pitched to them as the equivalent of a combat assignment – which would be wrong to refuse – with the astronauts as Cold War warriors.

"We'd had that as part of our background, the competitive mode with the Soviet Union. While I was in test pilot school, I was writing a report on an airplane that just barely exceeded the speed of sound. One night, I looked up and saw the booster, not *Sputnik* – nobody really saw *Sputnik*, they saw the booster – flying by, and I said, that's doing Mach 25, what am I doing working on this slow airplane? I forgot about it, until I was ordered to Washington. Then I began thinking, maybe this is the way we should be going, not just sitting back waiting for something to happen – become part of it. It was a tough decision to make, because I realized I was going to lose a lot of opportunities. As a Naval officer, I was trained, essentially bred, to be a military aviator. I was a Naval officer on assignment, not an employee of NASA. But I made a decision that was apparently not retraceable, or it turned out that way anyway. By the time the second astronaut group arrived, they started sliding into the NASA family. Our first group, we didn't know what would happen: no one did. At the end of our NASA careers, it turned out that no one had a place for us in the military."

Schirra began working his way through the selection process, including the medical tests. He had a clue that he might get through when the doctors seemed very keen to have him undergo a minor throat operation as soon as possible.

"It was a tipoff by going for the operation, though I already knew I was being considered. They said, 'If you are willing to do this visit to Bethesda, which is an unusual place for a young Naval officer to go, we'd consider you for a candidate.' I had a node on my vocal cord. The doctor who operated said something to the effect of, 'I have never worked on somebody this junior, you must be going to the Moon or something!' By then my tongue was being held by forceps, I couldn't say a word!"

Schirra was indeed selected as one of the Mercury 7 astronauts, and he accepted the offer. The program he now found himself in was somewhat confusing to him – though amongst military test pilots, he was expected to look and act far more like a civilian.

"There were no civilians in the initial candidacy. As a result, I guess I had the opinion that only military were eligible. But then once we were selected, the word came down from Eisenhower, they will not wear uniforms. We were to do things without the appearance of being military. Eisenhower also fought to have all the spaceflights publically acknowledged and observed, in contrast to the Soviets being very secretive."

In one area, Schirra did already look more like a civilian – his haircut. The others all had military-style buzzcuts, but Schirra kept his hair longer. "I didn't feel I had to be a Marine! I kidded the others about that. I kept the same haircut."

Schirra found that he had much in common with the other Mercury astronauts. As well as being test pilots, they were all from similar, in some cases sheltered backgrounds, with a similar set of values.

"All of us were small-town boys. In my little town I knew the police chief, all the policemen, the firemen – they were surrogate fathers. If I ever did anything wrong, one of them would be on my case, and if they ever came to the door, I knew I was in deep trouble. Today, people look at a police car and get worried. They get annoyed when they have to get out of the way of a fire truck – it's terrible. The police now are being harassed all the time. It's just a very sad transition. Talking about a small town, I played English football – soccer – instead of American football, because we couldn't afford the equipment. We just had shoes and shorts and you were in. I went to high school in Englewood, New Jersey. When I was 17, driving home at night, I stopped in a roadside diner and was having some eggs, and bacon and toast, and there was this black guy – the first time I'd ever seen one. He was having bacon and eggs. I looked over, and after a while he said, 'What do you expect, some grits and watermelon?' And I laughed, and he laughed – it was the perfect introduction to a black man. That was the first time I'd even seen one. You can just imagine the expressions, but he handled it so beautifully."

Schirra's upbringing and military background had given him little experience to cope with the instant fame and attention of being an astronaut. It took some time to adjust to his new role in the limelight. Each of the new astronauts handled it differently.

"I learned to live with it, and even have fun with it. But in quieter times, we really wanted to get away from it. Occasionally, when I am eating in public, I'm about ready to cut into this gorgeous entree, and somebody comes up and says, 'Wally, will you please sign this autograph for my son, he'll kill me if I come home without it.' My fantasy was to say, 'Don't tell the little bastard you saw me!' You have to learn to be tolerant of it. Shepard wasn't until his later years, then he became more mellow. I guess I was more outgoing. Shepard was what we called the icy commander. He wasn't like that with any of us, but with strangers he was that way. Much like Neil Armstrong is today – you'll find he's a lovely guy, a very nice man, but pretty much a recluse. Neil has handled the fame very well. John Glenn craved the publicity. I think even John would admit that.

When he went into politics, that became pretty obvious! We all saw that he knew how to do public relations from that original press conference. We weren't prepared for that at all – we were all looking over, thinking, what is this guy saying? We finally did adapt."

The seven astronauts soon realized that they could be powerfully influential as a group. If they all agreed that something needed to be changed, and presented the idea to management as a unified group, it usually was changed.

"Walt Williams, our Mercury Operations Director, used the term 'séance' to refer to what we did. When we came out of there, we would have a unanimous opinion, a couple of black eyes, and a few bruised shoulders! But we did it very rarely, we didn't overuse it. It wasn't like in *The Right Stuff* movie – that was just for entertainment."

Schirra was already used to trusting his life to other test pilots when it came to big decisions. He has described test piloting and the space program as the ultimate peer programs – you had to rely on your fellow pilots all the time.

"I think that is part of the game. When a test pilot comes off a flight, there is typically another pilot who is going to take it up, and he believes in the debriefing. You don't keep something to yourself. Within NASA, there were lots of things that were not appropriate to bring out to the public, because the press did not handle it well half the time. But within ourselves, we'd tell it all."

While the group had this deep level of trust, they could also be fiercely competitive with each other, each trying to prove that they were the best at everything. No weaknesses would be admitted to. For example, Deke Slayton began scuba training without revealing that he could not swim.

"Deke Slayton was the best diver we had – he went right to the bottom! Gus Grissom and I had to pull him off the bottom, and help him tread water. In the movie *The Right Stuff* they showed him cavorting with some girls in a water tank in a bar – the last person you'd really have put in that tank would be Deke Slayton! He was a farm boy out of Wisconsin. A river or a cistern was about the nearest thing he saw with water. Our competition was like sibling rivalry almost, but we bonded completely, and forever. We are still very close – I am currently furious at Chris Kraft for giving Scott Carpenter such a hard time in his book. I'm giving Chris a hard time back, which I normally would not do."

Schirra was always careful to give the impression of being unfazed by anything, no matter how hard the training got. When a centrifuge Schirra was riding span out of control, and others were worried he would be injured, he emerged with a grin on his face.

"That story was a little overstated, but it was a very high G load. The centrifuge ran away, it lost its braking effect of stopping at about 13 or 14 G. I think we peaked at about 18. I am sure I blacked out finally, but apparently I resolved myself to live through it. It was something I didn't need to do, and I have fought the centrifuge ever since. When I visited Star City in Russia in about 1991, I couldn't believe they still had one. I told them, you guys don't need a centrifuge, they are a waste of time. Deke Slayton said the same thing to them when he was over there for Apollo-Soyuz in 1975."

The competition between the astronauts was to try and get that coveted first spaceflight. Schirra was devastated not to be the one picked.

"We all were, I'm afraid. That was a lot of competition we had. It was just one of those philosophies that gets to you. Each test pilot I know considers him, or herself – now that there are women – to be the very best. It's very demeaning to step down the ladder once in a while. You feel filtered out. But I really think I had a better flight because of the delay."

In April 1962, before Schirra had flown, some more future competition arrived, in the form of the second group of astronauts. Some of the original seven seemed resentful to the new arrivals, and reluctant to share the benefits. Schirra was the one who went out of his way to include them and make them and subsequent groups feel welcome – even being described as a 'mother hen' by one astronaut.

"I was told several times by those guys, thanks for bringing us in. Pete Conrad and Jim Lovell I knew well already as they had been classmates of mine at test pilot school – Class 20. All three of us went through those early Mercury tests together, to see which one of us would survive, and somehow I did! Pete and I water skied a lot back then, and he and I tied for second in our test pilot class. Lovell studied, so he became first! Lovell had some kind of anomaly with his liver at the time of those Mercury tests and Conrad, they felt, wasn't able to live alone in space, or endure in space. The shrinks pretty well screwed up on that one. He lived on a space station for a month and flew Gemini 5, a long duration flight! On flights like those, it's not like you can eat out! One of my favorite memories is a picture of my Mercury launch, and that second group, called The Nine, standing there looking at the launch. Pete's fingers were crossed. It is a precious picture."

After serving as a backup to Scott Carpenter, Schirra's Mercury launch came on October 3, 1962, atop an Atlas booster, and he took it very seriously indeed. He considered himself an engineer and pilot, not a poet, and went as far as to ignore the view as much as possible. Schirra was not about to say something nice for public relations purposes.

"This is the game that people play. I'll never forget alan Shepard, on the very first manned American flight, saying something to the effect of, what a beautiful view. I asked him later, did you see anything at all? He said, 'I couldn't see a damn thing through that periscope – but I had to say something nice!' Getting the job done was more my logic. I kind of adopted that philosophy."

Schirra named his spacecraft *Sigma 7*, symbolizing engineering precision, and had the opportunity to independently pilot the spacecraft more than any previous flight. He considers that mission the first true piloted orbital flight.

"Glenn was turned around automatically, and was in automatic mode a lot, as was Carpenter. They used up a lot of fuel. So during my flight I put it in what we call a drifting mode, and I moved it back into attitude hold mode with minimum fuel, then put it in what I called 'chimp mode.' That didn't go over too well, but that was the point! It was getting back at the people calling us 'spam in a can.'"

Mercury was a very successful program, but to achieve the objective of a lunar landing, there was still much to be done. A vital step towards making a Moon landing would be practicing rendezvous between spacecraft, and the Gemini spacecraft would be the one to do it. After serving as backup commander to Gus Grissom for the first Gemini flight in 1965, Schirra prepared for his second flight, commanding Gemini 6 and piloting it to a rendezvous.

"It was a crucial step. [Jim] McDivitt and [Ed] White on Gemini 4 had screwed up badly with rendezvous. Buzz Aldrin tried to get credit for our success, but he had almost screwed us up. Buzz had this academic mind, and realized that there were two ways of doing a rendezvous. One way – his way – was mathematically pure, but if you messed up a little bit, you'd really mess it up. I told him we were not going to do it that way. He was talking about perfection, and perfection is when you're docked – not doing the maneuver. So we looked at a different way – the right way of doing it. We spent a lot of time in simulators. Dean Grimm, a NASA engineer, was one of the guys I worked with. I am indebted to those kinds of people. Of course, [John] Young and [Gus] Grissom were our backups and they worked on it too – about five or six of us really spent a lot of time on it. McDivitt and White were more involved in doing their spacewalk and other things and didn't really understand what we were doing. They didn't really have time to do that. We were going to do it right."

Things did not go right at first. Schirra and his pilot, Tom Stafford, tried to launch aboard Gemini 6 on October 25, 1965, and rendezvous with an unmanned Agena rocket. However, the Agena exploded, and the launch was scrubbed. The bold decision was made to launch Gemini 7 first, quickly refurbish the pad, and launch Gemini 6 for a rendezvous of both manned spacecraft.

Gemini 7 launched successfully, and on December 12, Schirra and Stafford were ready to try launching again. But something went wrong. The Titan booster's engines ignited, but the rocket did not lift from the pad. Schirra could feel by the seat of his pants that they had not moved.

"I had heard the booster liftoff in Atlas, and this Titan didn't work exactly the same way, so I knew in milliseconds that something had gone wrong, that we had not lifted off. The rule was to eject, to punch us both out. That was kind of a mission rule, and that's another of those rules that was kind of a what-if. And the what-ifs are not all necessarily in a row, or in the proper sequence."

Had Schirra ejected from the spacecraft, it would have been damaged beyond repair, and the rendezvous would not have happened. Schirra chose not to follow that mission rule. Instead, he and Stafford sat through the tense moment, until it was safe to unstrap and exit the spacecraft. Because of that risky decision, they were ready to try again three days later. They launched successfully, and rendezvoused with the waiting Gemini 7.

"Did I tell you exactly what a rendezvous is? When a man looks across a street and sees a pretty girl, and waves at her, that's not a rendezvous, that's a passing acquaintance. When he walks across the street through the traffic and nibbles on her ear, that's a rendezvous!"

Gemini became Schirra's favorite flying vehicle, closest of all the things he flew to being the ultimate for a test pilot – the harmony of man and machine.

"I appreciated it the most. In Mercury, you couldn't translate. You could just change attitude. But you were actually flying it like a flying machine in Gemini. Apollo was just too big, like flying a big transport airplane, which fighter pilots don't really revere. Gemini was just about the right size – it was not much larger than Mercury, really, and it was optimized for both my right and left hand. I did translation with my left hand, which is a very delicate maneuver – a bit like when the Shuttle docks with a space station. My left hand was better than my right hand for that."

With the rendezvous accomplished, Schirra had made a vital contribution to the success of the upcoming Apollo program. In mid-1966, Schirra, Walt Cunningham and Donn Eisele were named as the crew for the second manned Apollo flight. It was to be an identical mission to the first Apollo flight, for which a crew had also been chosen – Gus Grissom, Ed White and Roger Chaffee. Schirra could not see the point of flying a repeat mission, and his arguing succeeded in getting his mission canceled. However, this also meant that his crew was without a mission, and they were placed as Grissom's backup crew.

"I had a prime flight scheduled, Apollo 2. Well, it could have been called Apollo 2. I'm not sure what the real number would have been, because the numbers were all changed later to honor Apollo 1. I convinced the NASA people it was a dumb flight, to do the same thing all over again, much like the second Mercury flight. We finally stopped doing that in Gemini, and I asked why we were doing it in Apollo, if we were in a hurry to get to the Moon and back. So they made us backups. I was furious. Having been a Mercury backup, then a Gemini backup – this was three backups, and that was too much. Cooper wasn't even a prime any more. We were being pushed around a little bit, and I didn't like that very much."

Francis French (*left*) with Walt Cunningham and Wally Schirra at a Burbank space show in 2004. (Photo: Francis French)

Walt Cunningham speculated in his autobiography that the changes in crew assignments were to allow Deke Slayton, who was grounded with a medical condition, a flight to command. Schirra would be essentially a 'caretaker commander,' standing in until Slayton received medical clearance. Schirra refutes the story.

"Walt might have said that, but I would never have done that. Walt has lots of little fanciful ideas like that, once in a while. My wife said about Walt, 'He's like a puppy dog, keep scratching him and he'll be nice.' I stopped scratching him, and boy, he got nasty! But he's all right, really."

There was something else happening that threatened the chances of this crew flying – it was becoming obvious within NASA that Donn Eisele was having an affair that might lead to a divorce. Deke Slayton had warned the astronauts that they were all "expendable," and any extramarital affairs were never to make it into the papers. The only astronaut who had filed for divorce, Duane Graveline, had been thrown out of NASA so fast that he never even appeared in his group's official photo. Schirra was aware of what was going on.

"Donn Eisele was already entranced with a girl at the time of our flight, Susie, who was messing around with him in those days, and he later married her. I made note of it, not to a great degree, but I made note of it. Some of the details were accurately shown in the *From the Earth to the Moon* TV series. Some of the wives didn't keep up with the program. It started breaking apart during the Apollo days. Eisele divorced his wife after our Apollo flight, and then the flood came, the dam broke. At first, Cooper was essentially living apart from his wife, when he came in, but they came back together, and finally divorced after he left the space program. Walt Cunningham divorced after he left NASA, and remarried. Dick Gordon had six kids – divorced, and remarried a nice lady. I was looking at a book *Astronauts and their Families* just the other day; it came out just before my Gemini flight. I was really shocked how few of those guys are married to those women anymore. Our kids today don't want to get married. Too many of their friends have been married and divorced already. They and all of their friends just don't believe in it, they don't feel very comfortable with the idea of getting married."

Schirra, on the other hand, is still married to Jo, the same person he was married to before he became an astronaut.

"We have managed to hang in for 55 years, which isn't bad. My wife says our marriage has lasted so long because I was away half the time!"

Schirra considered it a personal letdown to once again be a backup, but set to work with Grissom's crew on trying to solve the problems that were delaying the manufacture of the Apollo 1 spacecraft. The slipping schedule meant that Schirra and Grissom's crews spent most of 1966 traveling between manufacturing plants, keeping the Apollo program moving forward.

On January 26, 1967, Schirra and his crew did a full systems test of the Apollo 1 spacecraft while it sat on the pad. The next day, Grissom's crew entered the spacecraft for the next series of tests – and died in a fire. In May 1967, after the investigation of the fire had been carried out, Schirra and his crew were named as the new crew to fly the first manned Apollo mission.

The Apollo 1 fire, and the responsibility to successfully fly the next mission, brought a different attitude to Schirra's work. Being a test pilot, Schirra knew how to live with the loss of a close friend like Gus Grissom – but that did not mean he was going to allow the same kinds of mistakes to be made twice. The light-hearted, joking Schirra was gone for a while, replaced by a hard-nosed commander who demanded attention to every detail. There was one important person Schirra insisted would be working with him, if the next flight were to be a success.

"I arranged for Guenter Wendt to be our pad leader. He essentially had been working for McDonnell Douglas for Mercury and Gemini. After the Apollo 1 disaster I asked North American Rockwell to hire him as our Apollo pad leader. They said, 'Do you want a Barbie doll too?' or something like that. I said, 'I don't think you fellows understand where I am coming from this time. You screwed up. I want a good man on the pad.' He did all the Apollo flights after that – North American hired him."

Schirra also decided that, to show he was single-minded about the upcoming flight, named Apollo 7, he would announce that this would be his last spaceflight.

"I wasn't about to stay in NASA by then; I knew what it was. I had always believed that I worked with NASA, not for NASA. There's a big difference! By 1968, I saw a bureaucracy developing – the fun days were over. I made the commitment that I would leave NASA, and wasn't sure then that I might not even retire from the Navy. I had been gone from the Navy for over ten years, and had lost all those stepping stones, commands that would have entitled me to have been promoted to Rear Admiral. I eventually decided to retire from the Navy as well as leave NASA."

Once Schirra had made the decision to leave, he felt that it gave him far more freedom to make criticisms of the spacecraft design that he felt were necessary. He and his crew pushed the spacecraft engineers further than ever before in their detailed rebuilding of the Apollo spacecraft, and pulled no punches.

"It helped, and it worked, too! I'm afraid others didn't always like it – they didn't realize what a command was. Particularly Chris Kraft. He didn't make a big issue out of it, but he did say I was kind of grumpy. I wasn't grumpy; I was merely asserting my authority. The flight controllers felt like they had the right to the last word – they still do! But I was taking the risks. I have yet to hear of a flight controller killing himself by falling off his chair! They were younger men who had not really put themselves physically at risk. They could wear black armbands, but that wouldn't help me any. The result of it was, when we lost three men on the launch pad, I knew we were facing up to a real problem. I said, 'We are going to do this one right.' By then I had responsibility for two crewmen. Then you have the responsibility, much like the skipper of a submarine – it's your problem. You accept command. That's the way it goes – if you give me the ship, it's mine. Then I'll tell you what I'm going to do or not do, within the rules of the ship. That comes down from the Royal Navy, the authority invested in the commanding officer."

In the *From the Earth to the Moon* TV series dramatizing Schirra's Apollo flight, there was a scene in which Schirra is shown telling Slayton he will be leaving NASA, but hinting that he might stay if he were given command of a flight to the Moon. Schirra says that this was artistic license:

"That was overplayed, no. The rule had been established by then, that was a published rule, that he who commands an Apollo flight will not command a second one. And it turned out to be true. The only one who flew two was Stafford, who had Apollo 10, and Apollo-Soyuz, which doesn't really count. There were a lot of guys waiting in line. I could see that I was out of line already. If Cooper was already out of line, how the heck could I get back in again? Betty Grissom said that Gus was in line to land on the Moon – that's a bunch of hogwash. That was pretty well bent out of shape. Deke never said that. In contrast, Deke said that we of the original seven are done, there's a whole new crew now. That I even got that Apollo flight was unusual. The second group was brought in to go to the Moon. We were supposed to be out of there by then. It just turned out they needed me, so I stayed for the Apollo 7 flight. That was unique."

Schirra and his crew were now fighting to have control of many aspects of the planning for their mission. One battle that Schirra had was to have coffee aboard the spacecraft.

"In the real world, most Naval officers live on coffee. At that time I was very much into coffee, much more so than even now. The spacecraft fuel cells made hot water, 150 degree water. So you could reconstitute freeze-dried coffee easily. In fact, you drink coffee at about 120–125 degrees Fahrenheit. However, the psychologists said it's a stimulant, that's all you want it for, it has no caloric value. So I let them go without it for a day – I removed it from their meetings. That worked out very well! I got my point across, and we had coffee on the flight."

There was another fight that Schirra did not win, however, that was fought right up until the hour of launch of Apollo 7, on October 11, 1968. The couches that the crew would sit in during the mission were of the old Apollo 1 design, and a newer, safer version would not be ready in time for this flight. The old couches were not designed to adequately protect the crew should the spacecraft abort during launch and come down on land. Schirra had insisted the rules on wind conditions for launch be revised, so that there would be no launch if there was a chance the spacecraft would be blown back onto land during an abort. As the countdown for the first manned Apollo mission neared the launch point, it was becoming obvious to Schirra that the winds were outside of the safe margins agreed upon – but no one was planning to stop the countdown.

"That was not the time to launch that day, and I didn't want to. Those were the wrong conditions, that violated mission rules. They broke the mission rule that we had established, that said we were not to launch under those conditions. We were not to launch if the wind was to blow us back over the beach, which would then force a land landing if we had to abort. That would essentially have been a death penalty. The winds on launch day were such that they would have blown us back over the beach. There was no problem about which day we launched. It was really a case of, someone wanted to go. I fought that, until I became rather difficult, and I finally yielded, with great concern. I conceded when we got to about T minus an hour and counting, when I realized that this could be a hard one to redo. But they were the ones who should have called the shutdown, not I. I tried to play it light. We launched because everything else seemed to be in good shape, it was one of those things, you say, okay, we'll take a go. I was furious. It was not as jocular as it looked on *From the Earth to the Moon*."

The unfortunate incident set the tone for relations between the spacecraft and Mission Control for the whole flight. The mission was an extremely successful test flight of the Apollo spacecraft, which allowed the next mission to be the first to leave Earth orbit and travel to the Moon. Everything that happened was done on Schirra's terms, however, as he exercised his full command of the spacecraft.

"I said, okay, if you're going to be violating rules, guess what I'm going to be doing! We're going to judge these rules from now on. If you are going to break that rule and not give me a chance – then I am going to break some of the rules that you have given me problems with! I didn't want to do things that hadn't been tested in their proper sequence. We were to test the circuit of the television set prior to using it. That was scheduled for one day, then the next day we were to turn it on. But the day they requested us to play games with the television, I was trying to do a rendezvous with the booster. I didn't want to mix that up with something else that was not important. They wanted the television on a particular day, and it wasn't scheduled for that day. I said we'll put it on tomorrow. That made sense to me – but not to them. Another real problem was over putting our helmets on for re-entry, because we all had severe head colds. They couldn't come up there and make us. Houston, you have a problem! Apollo was a big, unwieldy vehicle. I had a problem with the flight controllers over that. I said, 'I am tired of changing attitude up here, we are being affected by the atmosphere.' They said, 'What do you mean?' It was such a large vehicle, it would try to fare its way like an airplane. There's a name for it, I forget the name for it, causing the spacecraft to try and get the trimmest attitude. They didn't ever anticipate that, and there we were, very sensitive to anything that caused the vehicle to move. You don't read that on instruments, you have to feel it. But having said all that, I felt a lot for the flight controllers, and worked with them, not against them."

Following his successful flight, Schirra resigned from NASA in July, 1969. Fifteen days later, Apollo 11 was on the way to the Moon, and Schirra partnered Walter Cronkite during the television coverage of the mission. Looking back now, Schirra says he is still undecided as to whether Apollo was a stunt or not.

"In essence, as I look back on it, the timeframe was that we had a real beautiful Cold War going on. The challenge was that we had made a mess, or [President John] Kennedy had made a mess, in Cuba at the Bay of Pigs. In retrospect it seems that he had to do something to look good. The Apollo program concept of going to the Moon and back before the decade was out was quite a goal, which we all accepted, because we all loved the man. He was the only young, committed President we've ever had. We've lost that kind of commitment since. And yet, in fact, if you think of the inherent risk that we had, and the amount of effort the country went to – because just about everybody went to work on it – it's quite apparent that we were somewhat set up, I would say. We should have done it, but we didn't need to do it in that big a hurry. It would have been a much better program to have real piloted vehicles all the way through. In fact, von Braun made quite an issue at one point about having an Earth orbit rendezvous. Then going from Earth orbit with the vehicle to the Moon, then back to Earth orbit rendezvous. Guess what that is – International Space Station! We wouldn't have got there so quick, but we'd have something left for it. The word 'stunt' also applies in the sense that once we did it, everybody forgot about it. When I was broadcasting the Apollo missions with Walter Cronkite, the audience dropped tremendously, to a point where we couldn't even get airtime at the end. So when I look back on it in retrospect, I think of it that way."

Schirra wrote his autobiography, *Schirra's Space*, in 1988, and outlined how he would have liked to have seen the space program develop – full use of the space shuttle, a space station, and a space tug. Thirteen years later, things have not turned out as he hoped.

"The station has not met the big goals as a scientific station, which it was supposed to be. If you bring a tourist in there, what is he going to do to the zero-G environment if he bumps into something? I had been on a review of the commercial use of space back in the early 80's. The Secretary of the Interior had formed a board of advisors for that particular project. We had senior officers from major corporations. We all concluded that there was no way you could make money in space. One of the senior officers, a company chairman, was asked, 'Would you put your stuff on these flights for science evaluation if you had to pay your way?' He said, 'No way, I couldn't afford to do that.' That pretty well makes the point. The space station was developed as a result, to bring the price of experimentation down to reality. They're not going to do it the way they are doing it now.

"I don't think the space station will ever do anything for exploration. I think the idea of putting people up there for a year or more is the only way you will get anywhere near the exploration concept. Now they are just getting started – not the time to think about going to Mars at all. If anything, NASA should start thinking about this planet, and creating a much more efficient booster. We had a project called NERVA [Nuclear Engine for Rocket Vehicle Application] back in the sixties. That was dropped – and it might today be just the way to start doing things.

"NASA has changed since Kennedy's day, too. A lot of people have become older, and left. But attitudes have changed, too. One of my rules was, always keep the door open. Come in anytime and talk about something. Recently, the doors have been closing rather regularly. Now you have to fight the bureaucracy – a bureaucracy is a closed-door organization. Send a memo instead. The excitement we had in the early days was the fun of working on something new, and challenging, and devoting a lot of time to it.

"That's like when people have asked if I'd like to go in the Shuttle – I said you don't get to fly it, except for landing, which I'd love to do. The rest of it is just boring holes, which I did – I've done all that. I wouldn't go unless I could command it, and that would take two to three years of training. I wouldn't want to spend that much time. I wouldn't do what Glenn did. I already have more flight time than Glenn has even now! When they asked me if I was jealous of John's shuttle flight, I said, 'No, I'm not that old.' I don't need the flight time, I have 300 hours in space; he had five! I too would have done anything to get out of the U.S. Senate! I used to do that to Shepard too, kidding him that it took me longer to get down on each of my three flights than he was up there on his first one. We had so much fun teasing. I miss him so much; we teased all the time."

....

A few weeks after we had lunch, I ran into Schirra again at a party. Over the noise in the room, I asked him if he was enjoying himself. He loudly exclaimed, "No!"

Surprised, I asked him to elaborate and tell me why not, as it seemed a very enjoyable party.

"Oh," he grinned, "I thought you asked if I was *behaving* myself!"

Such is Wally Schirra.

Appendix 4

USN Capt. Wally Schirra's C.V.

Education:

Newark College of Engineering (N.J.I.T)	1941
U.S. Naval Academy	1942–1945 (B.S.)
Safety Officers School (U.S.C.)	1957
U.S. Navy Test Pilot School (N.A.T.C.)	1958
NASA Astronaut Training	1959–1969
Honorary Doctorate in Astronautical Engineering, Lafayette College	1969
Honorary Doctorate in Science, U.S.C.	1969
Honorary Doctorate in Astronautics, N.J.I.T.	1969
Trustee, Detroit Institute of Technology	1969–1976
Advisor, Colorado State University	1977–1982
Trustee, National College, South Dakota	1983–1987

Business Experience:

Director, Imperial American (Oil & Gas)	1967–1969
President, Regency Investors (Leasing)	1969–1970
Founder, Environmental Control Co. (ECCO)	1970–1973
Director, J.D. Jewel (Chicken Comp.)	1971–1973
Director, First National Bank, Englewood, CO	1971–1978
Belgian Consulate for Colorado and New Mexico	1971–1984
Director, V.P., Chairman, Sernco	1973–1974
Director, Rocky Mountain Airlines	1973–1984
Director, Carlsberg Oil & Gas	1974–1975
V.P., Johns-Manville Sales Corp., Denver, CO	1975–1977
Director, Advertising Unlimited, Sleepy Eye, MN	1978–1987
Director, Electromedics, Denver, CO	1979–1985
President, Prometheus Systems, Inc.	1980–1981
Director, Finalco (Leasing Co.), Mclean, VA	1983–1988
Director, Cherokee Data Systems, Boulder, CO	1984–1986
Director, Net Air Int., Van Nuys, CA	1982–1989
Director, Kimberly-Clark, Neenah, WI	1983–1991
Independent Consultant, Schirra Enterprises	1979–2007
Director, Zero Plus Telecommunications, Inc., Campbell, CA	1986–2007

C. Burgess, *Sigma 7*, Springer Praxis Books, DOI 10.1007/978-3-319-27983-1

Civic Activities:

Advisory Board/Council, U.S. National Parks (Interior)	1973–1985
Director, Denver Organizing Committee for 1976 Olympics	1973–1975
Advisor, Flight for Life, Mercy Hospital, Denver, Co.	1978–1986
Trustee, Colorado Outward Bound School (COB)	1970–1974
COB Regional Trustee	1988–2007
Advisory Board, International "Up With People"	1976–2007
Trustee, "Give Kids the World" Foundation	1980–2007
Founder/Director, Mercury Seven Foundation	1982–2007
Director, San Diego Air & Space Museum (later emeritus director)	1984–2002
Trustee, Scripps Aquarium Director	1985–2007
Ocean Foundation (became trustee)	1985–1990
Trustee, Hubbs/Seaworld Research Institute	1990–2007
Sharps Hospital, Foundation Board, San Diego	1988–1998
International Council, the Salk Institute, La Jolla, Ca.	1989–1998

Organizations:

33 Degree Mason	
Society of Experimental Test Pilots (SETP: Fellow)	1958–2007
American Astronautical Society (Fellow)	1960–2007
Explorers Club (Fellow)	1965–2007
Makai Country Club, Kauai, Hawaii	1971–2007
Rancho Santa Fe Tennis Club	1985–1995
San Diego Yacht Club	1987–2007
Charlie Russell Riders, Charter Member	1985–2007
Rancheros Visitadores, Member	1989–2001
Desert Caballeros, Member	1989–1998
Durango Mountain Caballeros, Member	1989–2000
Q.E.D., San Diego	1989–2007
The Golden Aviators (Naval Aviators)	1989–2007

Show Business Experience:

Anchor Man, All CBS/Cronkite Space Coverage	1969–1975
National Spokesman, Assn. of American Railroads	1969–1972
National Spokesman, V.P. National Systems	1972–1973
Host/Moderator, CBS, Smithsonian (2 films)	1970–1971
Central Figure, Dartnell Corp. "Take Command"	1970
Central Figure, Alaska Pipe Line Film	1973
Central Figure, El Paso Co., LNG Documentary Film	1978
National Spokesman, Midex (Burglar Alarms)	1977–1981
National Spokesman, Aztec (Heaters)	1978–1980
Host/Moderator TV Talk Show SCOPE (KAO TV)	1978–1980
National Spokesman, Realty World	1979–1980
National Spokesman, Sentry Safe	1979–1985
National Spokesman, Electromedics	1979–1985
National Spokesman, Actifed (Burroughs Wellcome)	1983–1990
Guest Various Talk Shows	1969–2007

Honors, Inductions & Awards:

U.S. Navy Distinguished Service Medal
3 × Distinguished Flying Cross
3 × Air Medals
3 × NASA Distinguished Service Medals [Apollo 7 DSM awarded posthumously, Oct. 2008]
2 × NASA Exceptional Service Medals
Navy Astronaut Wings
Robert J. Collier Trophy, 1962
Kitty Hawk Award
Harmon Trophy, 1966 (Gemini VI & VII crews)
Honorary Command Pilot, Philippine Air Force, 1966
SETP Iven C. Kincheloe Award, 1963
Great American Award
Golden Key award
Haley Astronautic Award
Aerospace Hall of Fame
International Aviation Hall of Fame, San Diego, CA
New Jersey Aviation Hall of Fame
International Space Hall of Fame, Alamogordo, NM
National Aviation Hall of Fame, Dayton, OH
Wall of Honor – Naval Aviation Museum, Pensacola, FL
Astronaut Hall of Fame, Titusville, FL
Launching of USNS Wally Schirra (T-AKE 8), March 2009, San Diego, CA
Inducted into the New Jersey Hall of Fame, May 2010, East Rutherford, NJ

Appendix 5

Wally Schirra in popular culture

Wally Schirra has been portrayed in one movie and in two television mini-series. In the highly popular 1983 movie *The Right Stuff*, based on the best-selling book of that name by author Tom Wolfe, he was played by character actor Lance Henriksen, although his portrayal was unfortunately a lesser role in the epic movie. Schirra liked the book a lot, but expressed disappointment and dislike for the movie, and he never forgave the producers for portraying his friend Gus Grissom as a bungling sort of coward, which was way out of line.

"It was the best book on space," he told Hollywood reporter Vernon Scott, "but the movie was distorted and warped. An actor named Lance Henriksen played me. He said he played me introverted, quiet, and retiring. My God, that's not me. Gordon Cooper *was* like that – yet they made him a hot dog. And there were gross distortions. They portrayed Chuck Yeager as my mentor. Hell, I never met him until after the movie. We're friends now, but we weren't then. All the astronauts hated [the movie]. We called it *Animal House in Space*. The sad part is the truth is truly exciting adventure without being hoked up."[1]

In the Ron Howard-directed 1995 movie *Apollo 13*, there is a short sequence that shows Wally Schirra and Walter Cronkite on TV commentating on the landing on the Moon by Apollo 11. That delighted Schirra, because it made him the only astronaut to appear as himself in the movie.

The 1998 HBO television series *From the Earth to the Moon*, co-produced by Ron Howard, Brian Grazer, Tom Hanks and Michael Bostick, had the part of Wally Schirra played in one of the twelve episodes ("We Have Cleared the Tower") by Mark Harmon, who later starred in the TV series *NCIS*. The series, narrated by Tom Hanks, was based on the book of the same name by Andrew Chaikin.

C. Burgess, *Sigma 7*, Springer Praxis Books, DOI 10.1007/978-3-319-27983-1

The three actors who have portrayed Wally Schirra on film and television. From left: Lance Henriksen, Mark Harmon, and Aaron McCusker. (Photos: Wikipedia)

Another mini-series aired in 2014, based on the book *The Astronaut Wives Club* by Lily Koppel, in which Wally Schirra was played by Aaron McCusker, an actor from Northern Ireland.

Several years after his flight aboard Apollo 7, and due to the notoriety associated with his well-reported head cold on that space mission, Schirra appeared in television commercials for the pharmaceutical product Actifed, which was actually based upon a cold medicine that was developed for the Apollo program and prescribed by the flight surgeons. In a 1991 speech that he gave at the Seal Beach Naval Weapons Station in California, Schirra made a wry observation on those commercials.

"We [the astronauts] were told we were flying a capsule. I didn't like the word capsule. Capsule connotes something very small that you swallow or something you hang out in, but we really didn't like that, so we tried to call it a spacecraft. Can you imagine years later I was pitching a cold remedy made in capsule form? I couldn't believe I did that."[2]

Schirra was also involved in several writing ventures over the years. In the 1962 astronaut book *We Seven*, he was responsible for putting together three chapters: "A Polyp in Time," "Some Seances in the Room," and "Our Cozy Cocoon." In 1988 he teamed up with freelance writer Richard Billings to write his autobiographical memoir *Schirra's Space*. Seven years later, he participated in the book *Wildcats to Tomcats: The Tailhook Navy* with co-author Barrett Tillman and fellow Navy Captains Richard L. ("Zeke") Cormier and Phil Wood. In 2005 he joined long-time friend and the first director of Huntsville's U.S. Space & Rocket Center, Ed Buckbee, in co-authoring a reflective book on the early days of the space program, *The Real Space Cowboys*. His final participation in a book project came about when he was interviewed extensively about his life, career, and accomplishments by Francis French and the author of this book for the 2007 *Outward Odyssey* companion books *Into That Silent Sea* and *In the Shadow of the Moon*.[3]

In a joke book promo, Schirra holds up a card similar to one that was displayed during a television transmission from Apollo 7. (Photo: Francis French)

ARE YOU AN ASTRONAUT?

REFERENCES

1. Vernon Scott, article "Schirra debunks notions about astronauts," *The Tribune* newspaper, San Diego, CA, 9 May 1985, pg. D-12
2. Wally Schirra speech, Seal Beach Naval Weapons Station, CA, 1991. Available online at: *https://www.youtube.com/watch?v=Fr4QSKNnOxM*
3. Information from *Wally Schirra* entry at: *https://en.wikipedia.org/wiki/Wally_Schirra*

About the author

Australian author Colin Burgess grew up in Sydney's southern suburbs where he and his wife Patricia still live. They have two grown sons, two grandsons and a granddaughter.

His working life began in the wages department of a major Sydney afternoon newspaper, where he first picked up his writing bug, and later as a sales representative for a precious metals company. He subsequently joined Qantas Airways as a passenger handling agent in 1970 and two years later transferred to the airline's cabin crew. He would retire from Qantas as an onboard Flight Service Director/Customer Service Manager in 2002, after 32 years' service.

During that period, several of his books were published about the Australian prisoner-of-war experience, as well as the first of his biographical books on space explorers such as Australian payload specialist Dr. Paul Scully-Power and *Challenger* teacher Christa McAuliffe. He has also written extensively on spaceflight subjects for astronomy and space-related magazines in Australia, the United Kingdom, and the Unites States.

In 2003 the University of Nebraska Press appointed him Series Editor for their ongoing *Outward Odyssey* series of books detailing the entire social history of space exploration, and he was involved in co-writing three of these volumes. His first Springer-Praxis book, *NASA's Scientist-Astronauts*, co-authored with British-based space historian David J. Shayler, was released in 2007. *Sigma 7* will be his eleventh title with Springer-Praxis, for whom he is currently researching further books for future publication. He regularly attends astronaut functions in the United States and is well known to many of the pioneering space explorers, allowing him to conduct personal interviews for these books.

C. Burgess, *Sigma 7*, Springer Praxis Books, DOI 10.1007/978-3-319-27983-1

Index

C. Burgess, *Sigma 7*, Springer Praxis Books, DOI 10.1007/978-3-319-27983-1

Printed in the United States
By Bookmasters